Green Energy and Technology

Hermann-Josef Wagner · Jyotirmay Mathur

Introduction to Wind Energy Systems

Basics, Technology and Operation

With 33 Figures and 6 Tables

 Springer

Prof.Dr. Hermann-Josef Wagner
Ruhr-Universität Bochum
Lehrstuhl für Energiesysteme
und Energiewirtschaft
Universtitätsstr. 150
44801 Bochum
Germany
lee@lee.rub.de

Dr.-Ing. Jyotirmay Mathur
Malaviya National Institute of
Technology
Dept. Mechanical Engineering
Jaipur-302017
JLN Marg
India
jyotirmay.mathur@gmail.com

ISBN 978-3-642-02022-3   e-ISBN 978-3-642-02023-0
DOI 10.1007/978-3-642-02023-0
Springer Heidelberg Dordrecht London New York

Springer Series in Green Energy and Technology ISSN 1865-3529

Library of Congress Control Number: 2009932563

The use of general descriptive names, registered names, trademarks, etc. in this publication does not
imply, even in the absence of a specific statement, that such names are exempt from the relevant protective
laws and regulations and therefore free for general use.

Cover design: WMXDesign GmbH, Heidelberg

Printed on acid-free paper

Springer is part of Springer Science+Business Media (www.springer.com)

Preface

In the past few decades, growth in the wind energy sector has been most phenomenal among all the renewable sources of energy. Consensus exists almost worldwide that for ensuring a sustainable future, wind energy can definitely play an important role.

The present book has been written to satisfy the interest of readers on wind energy converters. The authors have tried to strike a balance between a short book chapter and a very detailed book for subject experts. There were three prime reasons behind doing so: first, the field is quite inter-disciplinary and requires a simplified presentation for a person from non-parent discipline. The second reason for this short-version of a full book is that both authors have seen students and technically oriented people who were searching for this type of book on wind energy. The third reason and motivation for writing this book was to provide some initial information to people who are embarking on a career in the wind industry. It is this group of people that the present book is targeted at

This book presents the basic concepts of wind energy in its first two chapters. Chapter 3 deals with the physics and mechanics of operation. It describes the conversion of wind energy into rotation of turbine, and the critical parameters governing the efficiency of this conversion. Chapter 4 presents an overview of various parts and components of windmills with a blend of basics and recent advancements. Chapter 5 is dedicated to design considerations while selecting appropriate wind turbines for any site. Design options have been presented with their advantages and disadvantages. Chapter 6 is devoted to the utilization and control of operation of wind turbines. In this chapter, various parameters and methods for optimizing the performance are discussed. Chapter 7 presents the economics and financial issues associated with wind energy systems on the example of India and Germany.

Attempts have been made to include in this book technological advancements up to the beginning of the year 2009.

The authors thank Mrs. G. Schultz-Herzberger for controlling the English language of the manuscript. A special thank is given to Mrs. M. Koetter for her great help by typing and by controlling the text.

Authors also thank Prof. N.K. Bansal and Prof. R. Hanitsch for their suggestions. Support from DAAD for development of this book is also thankfully acknowledged.

The authors wish the readers a happy journey through the interesting field of wind energy.

Contents

List of Figures

List of Tables

About the Authors

Prof. Dr.-Ing. Hermann-Josef Wagner is postgraduate and doctorate in electrical and mechanical engineering from Technical University of Aachen, Germany. He is a Professor for Energy Systems and Energy Economics at the Ruhr-University of Bochum, Germany. He worked as a scientist for the Research Centre Juelich, for the German Parliament and for different universities. His relevant experiences are on the fields on energy systems analysis, renewable energies like wind energy and life cycle analysis, e-mail: lee@lee.ruhr-uni-bochum.de

Dr.-Ing. Jyotirmay Mathur is a mechanical engineer postgraduate in energy studies from Indian Institute of Technology, Delhi, India; and doctorate from University of Essen, Germany. He specializes in the areas of renewable energy systems, energy policy modelling and energy efficiency. Working as Reader in the Malaviya National Institute of Technology in Jaipur, Dr. Mathur has been founder coordinator of the postgraduate program in energy engineering at his institute. He is involved with several committees of national importance in India, e-mail: jyotirmay.mathur@gmail.com; jyotirmay@mnit.ac.in

Chapter 1
Wind Energy Today

1.1 Status

Wind energy is one of the oldest sources of energy used by mankind, comparable only to the use of animal force and biomass. Ancient cultures, dating back several thousand years, took advantage of wind energy to propel their sailing vessels. There are references to windmills relating to a Persian millwright in 644 AD and to windmills in Persia in 915 AD. These early wind energy converters were essential for pumping water and grinding cereals. In Europe, wind wheels were introduced around 1200 BC probably as an after-effect of the crusade to the orient. These European windmills were mainly used for grinding, except in the Netherlands where wind wheels supplied the power to pump river water to the land located below sea level. Between 1700 and 1800 AD the art of windmill construction reached its peak. The construction knowledge was relatively high and improved through trial and error. Later on, theories were developed, e.g. those by Euler, providing the tools to introduce new designs and, thus, to substantially improve the efficiency of energy conversion. Many windmills were built and operated in Denmark, England, Germany and the Netherlands during the eighteenth century. In 1750, the Netherlands alone had between 6,000 and 8,000 windmills in operation. The number of windmills in Germany has been estimated at about 18,000 in 1895, 11,400 in 1914, and between 4,000 and 5,000 in 1933.

Around the beginning of the twentieth century, windmills were further improved and the design of a multi-blade farm windmill originated in the USA. By the middle of the century, more than 6 million windmills were in operation in the USA. Worldwide, many of these wind wheels were used to produce mechanical power or as decentralized electricity suppliers on large farms. When the central electricity grid reached every farmhouse at the beginning of the twentieth century, the use of electricity produced by windmills rapidly decreased and the converters were taken out of operation as soon as the next repair job was due. In the nineteen fifties, pioneers like Huetter at the University of Stuttgart, Germany, took up developing and testing modern wind wheels again. Their design is quite different from the previous one; there are only two or three blades with very good aerodynamic parameters, able to rotate at high speed. Due to the high number of rotations, only a small generator is needed to produce electricity. Nevertheless, it was not possible to break

H.-J. Wagner, J. Mathur, *Introduction to Wind Energy Systems*, Green Energy and Technology, DOI 10.1007/978-3-642-02023-0_1, © Springer-Verlag Berlin Heidelberg 2009

Table 1.1   Use of wind energy world wide

Status of installed wind power	Rated capacity 1.1.2009 (MW)	Share worldwide (%)
USA	25,200	21
Germany	23,900	20
Spain	16,800	14
China	12,200	10
India	9,600	8
Italy	3,700	3
France	3,400	3
UK	3,200	3
Denmark	3,100	2
Portugal	2,800	2
Remaining countries	16,700	14
Total	120,600	100

even economically selling wind generated electricity in the fifties and sixties of the twentieth century. In the aftermath of the so called oil crisis in the seventies, there was a surge to enforce the development and marketing of wind wheels, especially in the USA, Denmark and Germany. This was based on the understanding that ultimately, additional energy sources emitting less pollution would be necessary. Due to favorable tax regulations in the eighties, about 12,000 wind converters supplying power ranging from 20 kW to about 200 kW were installed in California. In Europe, a lot of tax money was spent on the development of bigger wind converters and on marketing them. Now, in 2009, worldwide more than 100,000 windmills of about 120 GW installed power generation capacity are in operation, as shown in Table 1.1.

Germany had the leadership until 2007. Then USA has taken over the leadership. On the market are converters with a capacity up to 6 MW.

The market introduction of wind energy is proceeding in industrialized countries as well as in development countries like e.g. India. The Indian wind energy sector had an installed capacity of 9,600 MW as end of 2008. In terms of wind power installed capacity, India is ranked 5th in the World. Today, India is a major player in the global wind energy market. Still, the total potential of wind energy in the Indian country is far from exhausted. With the current level of technology, the "onshore" potential for utilization of wind energy for electricity generation is estimated of about 48,000 MW.

Up to the year 2008, the top five countries in terms of installed capacity are the USA, Germany, Spain, China, and India.

1.2 Advantages and Disadvantages of Wind Energy Systems

Wind energy offers many advantages, which explains why it is the fastest-growing energy source in the world. Research efforts are aimed at addressing the challenges to increase the use of wind energy.

1.2.1 Advantages

- Wind energy systems are energized by the naturally flowing wind, therefore it can be considered as a clean source of energy. Wind energy does not pollute the air like power plants that rely on combustion of fossil fuels, such as coal or natural gas. Wind turbines do not produce atmospheric emissions that cause acid rain or greenhouse gases.
- Wind energy is available as a domestic source of energy in many countries worldwide and not confined to only few countries, as in case of oil.
- Wind energy is one of the lowest-priced renewable energy technologies available today.
- Wind turbines can also be built on farms or ranches, thus benefiting the economy in rural areas, where most of the best wind sites are found. Farmers and ranchers can continue to use their land because the wind turbines use only a small fraction of the land. Wind power plant owners make rent payments to the farmer or rancher for the use of the land.

1.2.2 Disadvantages

- Wind power has to compete with conventional power generation sources on a cost basis. Depending on the wind profile at the site, the wind farm may or may not be as cost competitive as a fossil fuel based power plant. Even though the cost of wind power has decreased in the past 10 years, the technology requires a higher initial investment than fossil-fueled solutions for power supply.
- The major challenge to using wind as a source of power is that the wind is intermittent and it does not always blow when electricity is needed. Wind energy cannot be stored; and not all winds can be harnessed to meet the timing of electricity demands. The option of energy storage in battery banks is much beyond economically feasible limits for large wind turbines.
- Good wind sites are often located in remote locations, far from cities where the electricity is needed. In developing countries, there is always the extra cost of laying grid for connecting remote wind farms to the supply network.
- Wind resource development may compete with other uses for the land and those alternative uses may be more highly valued than electricity generation.
- Although wind power plants have relatively little impact on the environment compared to other conventional power plants, there is some concern over the noise produced by the rotor blades, and aesthetic (visual) impacts. Most of these problems have been resolved or greatly reduced through technological development or by properly siting wind plants.

1.3 Different Types of Wind Energy Converters: An Overview

Today, various types of wind energy converters are in operation (Fig. 1.1 gives an overview). The most common device is the horizontal axis converter. This converter consists of only a few aerodynamically optimized rotor blades, which for the

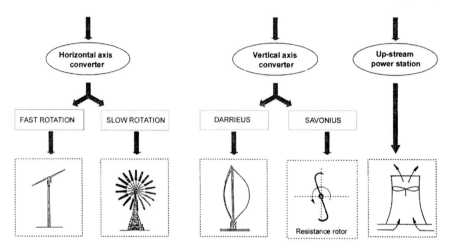

Fig. 1.1 Overview of different types of wind energy converters

purpose of regulation can usually be turned about their long axis (Pitch-regulation). Another cheaper way of regulation consists in designing the blades in such a way that the air streaming along the blades will go into turbulence at a certain speed (Stall-Regulation). These converters can deliver power ranging from 10 kW to some MW. The largest converter on the European market has a power of 6 MW. The efficiency of this type of wind energy converters in comparison with other types of windmills is very high. Another conventional (older) type of horizontal axis rotor is the multi-blade wind energy converter. It was first built about 100 years ago. Such windmills have a high starting torque which makes them suitable for driving mechanical water pumps. The number of rotations is low, and the blades are made from simple sheets with an easy geometry. For pumping water, a rotation regulating system is not necessary, but there is a mechanical safety system installed to protect the converter against storm damage. The rotor is turned in the direction of the wind by using a so called wind-sheet in leeward direction. The mechanical stability of such "slow speed converters" is very high; some have had operation periods of more than 50 years. In order to increase the number of rotations, this type of converter had been improved and equipped with aerodynamically more efficient blades facilitating the production of electricity, where the area of a blade is smaller.

A third type of converter is known as DARRIEUS, a vertical axis construction. Their advantage is that they do not depend on the direction of the wind. To start, they need because of their low starting torque the help of a generator working as a motor or the help of a SAVONIUS rotor installed on top of the vertical axis. In the 1980s, a reasonable number of Darrieus-converters were installed in California, but a further expansion into the higher power range and worldwide has not taken place. One reason may be that wind velocity increases significantly with height, making horizontal axis wheels on towers more economical. A modification of the Darrieus rotor is in the form of H-rotor; there are more than 30 installations of H-rotors worldwide but all of them are below the capacity of 300 kW. Prototype testing was successfully completed for this type but the commercial stage is yet to be seen. The SAVONIUS rotor is used as a measurement device especially for wind velocity; it is used for power production for very small capacities under 100 W.

The last technique to deal with is known as Up-Stream-Power-Station or thermal tower. In principle, it can be regarded as a mix between a wind converter and a solar collector. The top of a narrow, high tower contains a wind wheel on a vertical axis driven by the rising warm air. A solar collector installed around the foot of the tower heats up the air. The design of the collector is simple; a transparent plastic foil is fixed several meters above the ground in a circle around the tower. Therefore, the station needs a lot of space and the tower has to be very high. Such a system has a very poor efficiency, only about 1%. The advantage of such a design is its technical simplicity, which may enable developing countries to construct it by themselves. Worldwide, only one Up-Stream-Power-Station, designed by a German company, has been built so far. For some years it worked satisfactorily at the location of Manzanares in Spain, but in the mid eighties it was destroyed by bad weather. This station had an electrical power of 50 kW, the tower was about 200 m high and the collector had a diameter of approximately the same size. Another Up-Stream-Power-Station with an electrical performance of about 200 MW was planned in India. Other feasibility studies have also been conducted in Australia and Namibia but no project of this type has been realized.

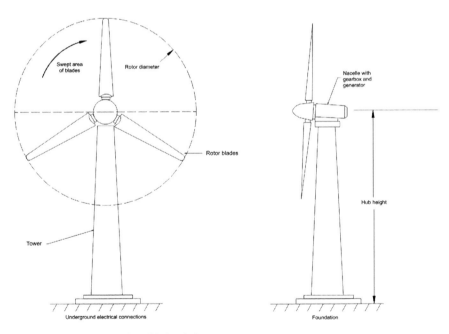

Fig. 1.2   Horizontal axis three blade wind energy converter

Throughout this book, the three terms "wind energy converters", "windmills" and "wind turbines" have been used quite interchangeably. The first term is the technical name of the system, whereas the other two are popularly used terms. Figure 1.2 shows the most popular type of horizontal axis three blade wind energy converter for generating electricity, worldwide. This figure shows the front and side views of a three blade horizontal axis wind energy converter. Details about its major parts and their working are discussed in subsequent chapters.

Chapter 2
Wind: Origin and Local Effects

All renewable energy (except tidal and geothermal power), and even the energy in fossil fuels, ultimately comes from the sun. About 1–2% of the energy coming from the sun is converted into wind energy. This chapter explains the cause of wind flow and factors that affect the flow pattern. Understanding these is necessary for selecting proper locations for wind turbines.

2.1 Origin and Global Availability

The regions around the equator, at 0° latitude, are heated more by the sun than the rest of the globe. Hot air is lighter than cold air; it will rise into the sky until it reaches approximately 10 km altitude and will spread to the North and the South. If the globe did not rotate, the air would simply arrive at the North Pole and the South Pole, sink down, and return to the equator.

Since the globe is rotating, any movement on the Northern hemisphere is diverted to the right, if we look at it from our own position on the ground. (In the Southern hemisphere it is bent to the left). This apparent bending force is known as the Coriolis force, named after the French mathematician Gustave Gaspard Coriolis.

In the Northern hemisphere, the wind tends to rotate counterclockwise as it approaches a low pressure area. In the Southern hemisphere, the wind rotates clockwise around low pressure areas.

The wind rises from the equator and moves north and south in the higher layers of the atmosphere. Around 30° latitude in both hemispheres the Coriolis force prevents the air from moving much farther. At this latitude, there is a high pressure area, as the air begins to sink down again. As the wind rises from the equator there is a low pressure area close to ground level attracting winds from the North and South. At the Poles, there is high pressure due to the cooling of the air. Table 2.1 shows the prevailing direction of global winds.

The prevailing wind directions are important when siting wind turbines, since one would obviously want to place them in areas with the least obstacles from the prevailing wind directions.

H.-J. Wagner, J. Mathur, *Introduction to Wind Energy Systems*, Green Energy and Technology, DOI 10.1007/978-3-642-02023-0_2, © Springer-Verlag Berlin Heidelberg 2009

Table 2.1 Prevailing global wind directions

Latitude	90–60°N	60–30°N	30–0°N	0–30°S	30–60°S	60–90°S
Direction	NE	SW	NE	SE	NW	SE

2.2 Local Effects on Wind Flow

Winds are very much influenced by the ground surface at altitudes up to 100 m. The winds are slowed down by the earth's surface roughness and obstacles. There may be significant differences between the direction of the global or geostrophic winds because of the earth's rotation (the Coriolis force), and the wind directions near the surface. Close to the surface of the earth, the following effects influence the flow pattern of wind:

(a) *Sea breezes.* Land masses are heated by the sun more quickly than the sea in the daytime. The air rises, flows out to the sea, and creates a low pressure at ground level which attracts the cool air from the sea. This is called a sea breeze. At nightfall there is often a period of calm when land and sea temperatures are equal. At night the wind blows in the opposite direction. The land breeze at night generally has lower wind speeds, because the temperature difference between land and sea is smaller at night.

The monsoon (specific period of the year when majority of rainfall occurs) in India and all of South-East Asia is in reality a large-scale form of the sea breeze and land breeze, varying in its direction between seasons, because land masses are heated or cooled more quickly than the sea.

(b) *Mountain breeze.* Mountain regions display many interesting weather patterns. One example is the valley wind which originates on South-facing slopes (North-facing in the Southern hemisphere). When the slopes and the neighboring air are heated, the density of the air decreases, and the air ascends towards the top following the surface of the slope. At night the wind direction is reversed, and turns into a down-slope wind. If the valley floor is sloped, the air may move down or up the valley, as a canyon wind. Winds flowing down the leeward sides of mountains can be quite powerful.

(c) *The wind rose.* It can be noticed that strong winds usually come from a particular direction. To show the information about the distributions of wind speeds, and the frequency of the varying wind directions, one may draw a so-called wind rose as shown in Fig. 2.1 on the basis of meteorological observations of wind speeds and wind directions.

The wind rose presents a summary of annual wind data. The circular space can e.g. be divided in 16 sectors representing major directions from which wind might come. The number of segments may be more but it makes the diagram difficult to read and interpret. The concentric circles having percent values represent the probability of wind coming from any particular direction. In Fig. 2.1, the polygon in north direction goes slightly ahead of the 10% circle, hence it can be concluded that at this location, the probability of wind coming from the North is roughly 12%. The

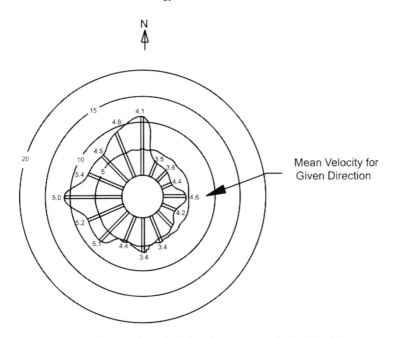

Fig. 2.1 Example of a wind rose (Annual all directions average velocity 4.3 m/s)

bars going up to the border of the polygon have a value tag with them. This value represents the mean velocity of wind when it comes from that particular direction. Again, in Fig. 2.1, the bar in the North direction says that when wind comes from the North it has an average velocity of 4.1 m/s. This graph shows that the strongest wind comes from the westward directions.

It should be noted that in popular terms, when we say that some site has "North wind", this means that wind is coming from the North and not going to the North.

Further to note is that wind patterns may vary from year to year, and the energy content may vary (typically by some 10%) from year to year, so it is best to have observations from several years to calculate a credible average. Planners of large wind parks will usually rely of local measurements, and use long-term meteorological observations from nearby weather stations to adjust their measurements to obtain a reliable long term average.

2.3 Attractive Locations for Wind Energy

In Europe, the area close to the North Sea has the strongest winds. In the extreme North, a wind velocity of the order of 10–11 m/s is available at a height of 50 m. This velocity is very attractive for installation wind machines. The Southern part of Europe also has wind energy potential but it has relatively less wind velocities, in the range of 3–8 m/s at the height of 50 m above the ground.

The Centre for Wind Energy Technology, India, has estimated a 48,000 MW potential of wind energy in India. This estimate is based on the current technologies and economics of wind energy. The Southern part of India is most attractive for wind energy. The wind energy map of India reveals that the maximum wind energy potential lies either in the Southern or Western part of India. In the states of Tamilnadu and Karnataka, there are sites having velocities in the range of 6–8 m/s at a 50 m height above the ground. But in the Western parts of India, close to the Arabian Sea as well as in the desert of Rajasthan, attractive wind energy locations also exist.

The wind data available from various agencies give a fairly good idea about attractive locations for wind energy installations. However, to get exact knowledge about certain locations, measurement of wind availability over several years would be appropriate. The reasons for this are explained in the next section.

2.4 Local Effects on Wind Flow

For the purpose of wind turbines, local wind is more important, since due to local effects, a site may have very low wind even if it is situated in a predominantly windy area. Major factors that govern local winds are, therefore, described in this section.

2.4.1 Roughness Length and Wind Shear

High above ground level the wind is influenced by the surface of the earth at all. In the lower layers of the atmosphere, however, wind speeds are affected by the friction against the surface of the earth. In the wind industry one distinguishes between the roughness of the terrain, the influence from obstacles, and the influence from the terrain contours, which is also called the orography of the area.

The more pronounced the roughness of the earth's surface, the more the wind will be slowed down. In the wind industry, wind conditions in a landscape are referred through roughness classes or roughness lengths. The term roughness length is the distance above ground level where the wind speed theoretically should be zero. A high roughness class of 3–4 refers to landscapes with many trees and buildings, while a sea surface is in roughness class 0. Concrete runways in airports are in roughness class 0.5.

2.4.2 Wind Speed Variability

The wind speed is always fluctuating, and thus the energy content of the wind is always changing. The variation depends on the weather and on local surface conditions and obstacles. Energy output from a wind turbine will vary as the wind varies,

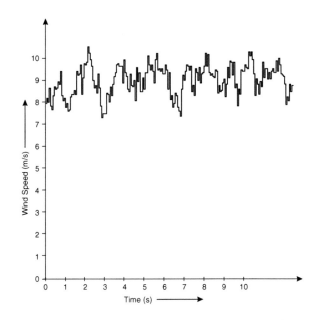

Fig. 2.2 Short term variability of the wind

although the most rapid variations will to some extent be compensated for by the inertia of the wind turbine rotor. Figure 2.2 shows short term variations in wind.

In most locations around the globe it is more windy during the daytime than at night. This variation is largely due to the fact that temperature differences, e.g. between the sea surface and the land surface, tend to be larger during the day than at night. The wind is also more turbulent and tends to change direction more frequently during the day than at night.

From the point of view of wind turbine owners, it is an advantage that most of the wind energy is produced during the daytime, since electricity consumption is higher than at night. Many power companies pay more for the electricity produced during the peak load hours of the day (when there is a shortage of cheap generating capacity).

At most locations, the wind may not be sufficient for producing power continuously for two to three days and sometimes even for one week.

2.4.3 Turbulence

It is normally experienced that hailstorms or thunderstorms in particular, are associated with frequent gusts of wind which both change speed and direction. In areas with a very uneven terrain surface, and behind obstacles such as buildings, a lot of turbulence is similarly created, with very irregular wind flows, often in whirls or vortexes in the neighborhood.

Turbulence decreases the possibility of using the energy in the wind effectively for a wind turbine. It also imposes more tear and wear on the wind turbine, as explained in the section on fatigue loads. Towers for wind turbines are usually made tall enough to avoid turbulence from the wind close to ground level.

2.4.4 Obstacles to Wind Flow

Obstacles to the wind such as buildings, trees, rock formations etc. can decrease wind speeds significantly, and they often create turbulence in their neighborhood. It can be seen in Fig. 2.3 that in case of typical wind flows around an obstacle, the turbulent zone may extend to some three times the height of the obstacle. The turbulence is more pronounced behind the obstacle than in front of it. Therefore, it is best to avoid major obstacles close to wind turbines, particularly if they are upwind in the prevailing wind direction, i.e. "in front of" the turbine.

However, every tower of a wind converter will work as such an obstacles and will have influence on the blade when it is coming before or behind the tower.

Obstacles will decrease the wind speed downstream. The decrease in wind speed depends on the porosity of the obstacle, i.e. how "open" the obstacle is. Porosity is defined as the open area divided by the total area of the object facing the wind. A building is obviously solid, and has no porosity, whereas a fairly open tree may let more than half of the wind through. In case of very dense trees, the porosity is less, say one third. The slowdown effect on the wind from an obstacle increases with the height and length of the obstacle. The effect is obviously more pronounced close to the obstacle, and close to the ground.

(a) *Reduction in obstacles with turbine hub height*. The higher a turbine is above the top of the obstacle, the less wind shade will be produced. The wind shade, however, may extend to up to five times the height of the obstacle at a certain distance. If the obstacle is taller than half the hub height, the results are more uncertain, because the detailed geometry of the obstacle, (e.g. differing slopes of the roof on buildings) will affect the result.

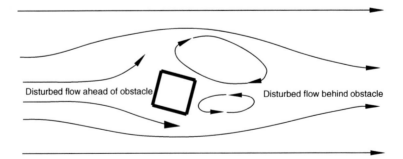

Fig. 2.3 Obstacle in wind flow

(b) *Distance between obstacle and turbine*. The distance between the obstacle and the turbine is very important for the shelter effect. In general, the shelter effect will decrease as one moves away from the obstacle, just like a smoke plume becomes diluted as we move away from a smokestack. In terrain with very low roughness (e.g. water surfaces), the effect of obstacles (e.g. an island) may be measurable up to 20 km away from the obstacle. If the turbine is closer to the obstacle than five times the obstacle height, the results will be more uncertain, because they will depend on the exact geometry of the obstacle.

2.4.5 The Wind Wake and Park Effect

Since a wind turbine generates electricity from the energy in the wind, the wind leaving the turbine must have a lower energy content than the wind arriving in front of the turbine. There will be a wake effect behind the turbine, i.e. a long trail of wind which is quite turbulent and slowed down, as compared to the wind arriving in front of the turbine. The expression wake is obviously derived from the wake behind a ship. Wind turbines in parks are usually spaced at least three rotor diameters from one another in order to avoid too much turbulence around the turbines downstream.

As a result of the wake effect, each wind turbine will slow down the wind behind it as it pulls energy out of the wind and converts it to electricity. Ideally, therefore turbines should be spaced as far apart as possible in the prevailing wind direction. On the other hand, land use and the cost of connecting wind turbines to the electrical grid would force to space them closer together.

As a guideline for wind park design turbines in wind parks are usually spaced somewhere between 5 and 9 rotor diameters apart in the prevailing wind direction, and between 3 and 5 diameters apart in the direction perpendicular to the prevailing winds. In Fig. 2.4, three rows of five turbines each are placed in a fairly typical pattern. The turbines (the dots) are placed 7 diameters apart in the prevailing wind direction, and 4 diameters apart in the direction perpendicular to the prevailing winds.

2.4.6 The Hill Effect and Tunnel Effect

(a) *Speed up effects: tunnel effect*. While walking between tall building, or in a narrow mountain pass, it can be noticed that wind velocity increases. The air becomes compressed on the windy side of the buildings or mountains, and its speed increases considerably between the obstacles to the wind. This is known as a "tunnel effect". So, even if the general wind speed in open terrain may be, say, 6 m/s, it can easily reach 9 m/s in a natural "tunnel". Placing a wind turbine in such a tunnel is one clever way of obtaining higher wind speeds than in the surrounding areas. To obtain a good tunnel effect the tunnel should be "softly" embedded in the landscape. In case the hills are very rough and uneven, there may be lots of turbulence in the area,

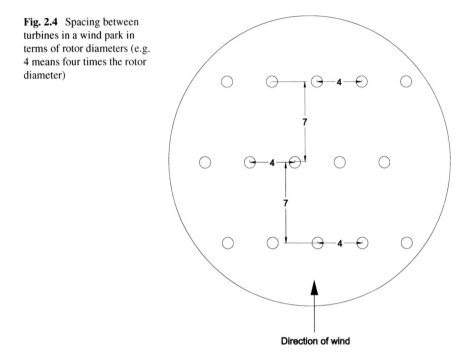

Fig. 2.4 Spacing between turbines in a wind park in terms of rotor diameters (e.g. 4 means four times the rotor diameter)

i.e. the wind will be whirling in a lot of different (and rapidly changing) directions. If there is much turbulence it may negate the wind speed advantage completely, and the changing winds may inflict a lot of useless tear and wear on the wind turbine.

(b) *The hill effect.* A common way of siting wind turbines is to place them on hills or ridges overlooking the surrounding landscape. In particular, it is always an advantage to have as wide a view as possible in the prevailing wind direction in the area. On hills, one may also experience that wind speeds are higher than in the surrounding area. Once again, this is due to the fact that the wind becomes compressed on the windy side of the hill, and once the air reaches the ridge it can expand again as its soars down into the low pressure area on the lee side of the hill.

2.5 Selecting a Turbine Site

Looking at nature itself is usually an excellent guide to finding a suitable wind turbine site. However, some typical considerations are as follows:

(a) *Wind conditions.* If there are trees and shrubs in the area, one may get a good clue about the prevailing wind direction. While moving along a rugged coastline, centuries of erosion which have worked in one particular direction can also be noticed. Meteorology data, ideally in terms of a wind rose calculated over 20–25 years is probably the best guide, but these data are rarely collected directly at the site, and there are many reasons to be careful about the use of meteorology data. Meteorologists collect wind data for weather forecasts and aviation, and that information is often used to assess the general wind conditions for wind energy in an area. Unless calculations are made which compensate for the local conditions under which the meteorology measurements were made, it is difficult to estimate wind conditions at a nearby site. In most cases using meteorology data directly will underestimate the true wind energy potential in an area.

If there are already wind turbines in the area, their production results are an excellent guide to local wind conditions. In countries like Denmark, Germany, Spain, and in the Southern part of India, where a large number of turbines are found scattered around the countryside, manufacturers can offer guaranteed production results on the basis of wind calculations made on the site.

(b) *Look for a view.* It is often preferred to have a view as wide and open as possible in the prevailing wind direction, and we would like to have as few obstacles and as low a roughness as possible in that same direction. If a rounded hill can be found to place the turbines, one may even get a speed up effect in the bargain.

(c) *Soil conditions and transportation facilities.* Both the feasibility of building foundations for the turbines and road construction to reach the site with heavy trucks must be taken into account with any wind turbine project. Due to the large size of equipment and machinery, these sometimes become bottlenecks for the installation of systems.

To sum up the coverage on wind, it may be said that for taking a macro level decision such as around which city the installation should be planned, the global wind availability and published wind data may be used. However, for an exact siting of wind energy system, a micro level analysis would be necessary due to large variations caused by the local geographical details. Only in case of very high turbines, e.g. 130 m high, where local disturbances have nearly no effect, it may not be necessary to go into micro level details. The exercise of finding an exact location for installation is termed as "micro-siting" which requires actual measurement of wind data at the site.

Chapter 3
Physics of Wind Energy

The basic principles of physics on which any wind turbine works are explained in this chapter. These concepts will be helpful in understanding the science and technology behind the operation and control of wind turbines in order to harvest maximum energy from the wind.

3.1 Energy Content in Wind

A wind turbine obtains its power input by converting the force of the wind into torque (turning force) acting on the rotor blades. The amount of energy which the wind transfers to the rotor depends on the density of the air, the rotor area, and the wind speed.

(a) *Density of air*. The kinetic energy of a moving body is proportional to its mass. The kinetic energy in the wind thus depends on the density of the air, i.e. its mass per unit of volume.

In other words, the "heavier" the air, the more energy is received by the turbine. At normal atmospheric pressure and at 15°C, the density of air is 1.225 kg/m^3, which increases to 1.293 kg/m^3 at 0°C and decreases to 1.164 kg/m^3 at 30°C. In addition to its dependence upon temperature, the density decreases slightly with increasing humidity. At high altitudes (in mountains), the air pressure is lower, and the air is less dense. It will be shown later in this chapter that energy proportionally changes with a variation in density of air.

(b) *Rotor area*. When a farmer tells how much land he is farming, he will usually state an area in terms of square meters or hectares or acres. With a wind turbine it is much the same story, though wind farming is done in a vertical area instead of a horizontal one. The area of the disc covered by the rotor (and wind speeds, of course), determines how much energy can be harvested over a year.

A typical 1,000 kW wind turbine has a rotor diameter of 54 m, i.e. a rotor area of some 2,300 m^2. The rotor area determines how much energy a wind turbine is able to harvest from the wind. Since the rotor area increases with the square of the rotor diameter, a turbine which is twice as large will receive 2^2, i.e. four times as much energy.

H.-J. Wagner, J. Mathur, *Introduction to Wind Energy Systems*, Green Energy and Technology, DOI 10.1007/978-3-642-02023-0_3, © Springer-Verlag Berlin Heidelberg 2009

Fig. 3.1 Power output
increases with the rotor
diameter and the swept rotor
area

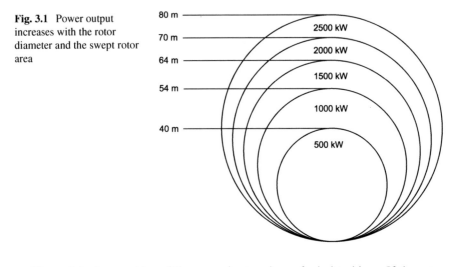

Figure 3.1 gives an idea of the normal rotor sizes of wind turbines: If the rotor diameter is doubled, one gets an area which is four times larger (two squared). This means that four times as much power output from the rotor will also be obtained.

Rotor diameters may vary somewhat from the figures given above, because many manufacturers optimize their machines to local wind conditions: A larger generator, of course, requires more power (i.e. strong winds) to turn at all. So if one installs a wind turbine in a low wind area, annual output will actually be maximized by using a fairly small generator for a given rotor size (or a larger rotor size for a given generator). The reason why more output is available from a relatively smaller generator in a low wind area is that the turbine will be running more hours during the year.

(c) *Wind velocity.* Considering an area A (e.g. swept area of blades) and applying a wind velocity v, the change in volume with respect to the length "l" is:

$$\Delta V = A \cdot \Delta l,$$
$$v = \frac{\Delta l}{\Delta t} \tag{3.1}$$
$$\Rightarrow \Delta V = A \cdot v \cdot \Delta t.$$

The energy in the wind is in the form of kinetic energy. Kinetic energy is characterized by the equation:

$$E = \frac{1}{2}mv^2 \tag{3.2}$$

The change in energy is proportional to the change in mass, where

$$m = V \cdot \rho_a \tag{3.3}$$

and ρ_a the specific density of the air. Therefore, substituting for V and m yields

$$E = \frac{1}{2} \cdot A \cdot \rho_a \cdot v^3 \cdot t. \tag{3.4}$$

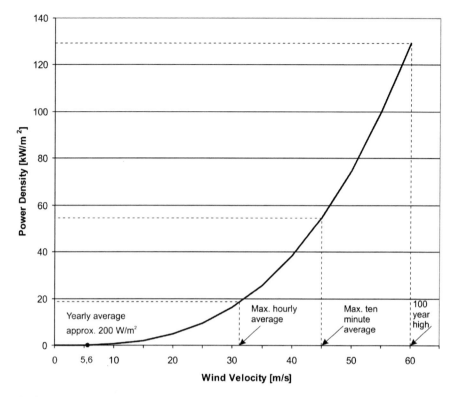

Fig. 3.2 Relationship between wind velocity and power of wind (wind speed for Germany)

From the previous equation it can be seen that the energy in the wind is proportional to the cube of the wind speed, v^3. The power P is defined as

$$P = \frac{E}{t} = \frac{1}{2} \cdot A \cdot \rho_a \cdot v^3 \tag{3.5}$$

Therefore, power in wind is proportional to v^3. From Fig. 3.2 it can be seen that the power output per m² of the rotor blade is not linearly proportional to the wind velocity, as proven in the theory above. This means that it is more profitable to place a wind energy converter in a location with occasional high winds than in a location where there is a constant low wind speed. Measurements at different places show that the distribution of wind velocity over the year can be approximated by a Weibull-equation. This means that at least about 2/3 of the produced electricity will be earned by the upper third of wind velocity. From a mechanical point of view, the power density range increases by one thousand for a variation of wind speed of factor 10, thus producing a construction limit problem. Therefore, wind energy converters are constructed to harness the power from wind speeds in the upper regions.

3.2 Energy Conversion at the Blade

Figure 3.3 shows the velocities and forces at the profile of a rotating blade. The blade itself moves with an average circumferential velocity u in the plane of the rotor. Rotation of blade provides a relative velocity of air with respect to the blade, which can be considered as additional wind (velocity vector u) working together with the actual wind to determine the rotational force on the blade. The wind flows perpendicular to the plane of the rotor, thus creating a resultant velocity vector w. This velocity is the relative approach or flow velocity of the rotor blade.

The two main forces acting on the rotor blade are the lift force F_A and the drag force F_R. The drag force acts parallel to the initial direction of movement and the lift force acts perpendicular to it. The lift force is the greater force in normal operating conditions and arises due to the unequal pressure distribution around an aerofoil

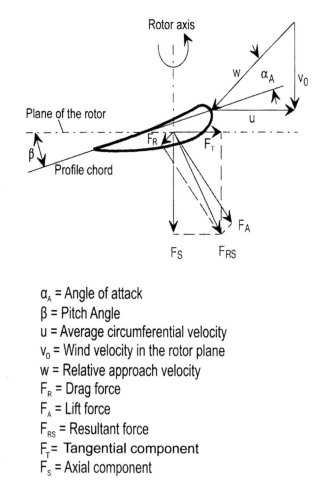

α_A = Angle of attack
β = Pitch Angle
u = Average circumferential velocity
v_0 = Wind velocity in the rotor plane
w = Relative approach velocity
F_R = Drag force
F_A = Lift force
F_{RS} = Resultant force
F_T = Tangential component
F_S = Axial component

Fig. 3.3 The velocities and forces acting on a blade

profile. The pressure on the upper surface is lower than that on the underside, therefore the air has a higher velocity when passing over the upper surface of the profile. The lift force is determined by the following formula:

$$F_A = \frac{1}{2} \cdot \rho_a \cdot c_A \cdot w^2 \cdot A, \tag{3.6}$$

where c_A is the lift force coefficient. The drag force is determined from a similar formula,

$$F_R = \frac{1}{2} \cdot \rho_a \cdot c_R \cdot w^2 \cdot A, \tag{3.7}$$

where c_R is the drag force coefficient, and is caused by air friction at the surface of the profile. The relationship between the two forces is given by the ratio E_G of their coefficients,

$$E_G = \frac{c_A}{c_R}. \tag{3.8}$$

This relation is at least given by the aerodynamic quality of the blade, by its design and its surface quality. It can be seen from Fig. 3.3 that the resultant force F_{RS} of the lift and drag forces can be divided into two components: the tangentially acting component F_T and the axially acting component F_S. It is the force F_T that causes the rotation of the rotor blade and makes power delivery possible.

3.3 Power Coefficients and Principles of Design

The wind turbine rotor must obviously slow down the wind as it captures its kinetic energy and converts it into rotational energy. This means that the wind will be moving slowly after leaving the rotor as compared to its movement before entering the rotor. Farther downstream, the turbulence in the wind will cause the slow wind behind the rotor to mix with the faster moving wind from the surrounding area. The wind shade behind the rotor will therefore gradually diminish as one moves away from the turbine. The factors governing the proportion of velocity ahead and behind the rotor that in turn govern the power output, are discussed in the following sub-sections:

3.3.1 Coefficient of Power c_p and Betz' Law

The question of how much of the wind energy can be transferred to the blade as mechanical energy has been answered by the Betz' law.

Betz' law states that only a maximum of 59.25% of the kinetic power in the wind can be converted to mechanical power using a wind turbine, the so called maximum power coefficient or Betz-Number. This number is not higher because the wind on the back side of the rotor must have a high enough velocity to move away and allow more wind through the plane of the rotor.

The relationship between the mechanical power of the rotor blade P_R and the power of wind P in the rotor area is given by the power coefficient c_p:

$$c_p = \frac{P_R}{P} \tag{3.9}$$

The power coefficient c_p can be interpreted as the efficiency between the rotor blade and the wind. The maximum power coefficient, the above mentioned Betz number, determined by the ratio $v_2/v_1 = 1/3$. Therefore, an ideal turbine will slow down the wind by 2/3 of its original speed.

3.3.2 Tip Speed Ratio

A wind energy converter is classified through the characteristic tip speed ratio λ_S. This is the ratio (as a scalar) of the circumferential velocity of the rotor at the end of the blade (maximum velocity u_e) and the wind velocity v_0 in front of the rotor blade:

$$\lambda_S = \frac{u_e}{v_0}. \tag{3.10}$$

The tip speed ratio has a strong influence on the efficiency of a wind energy converter (Fig. 3.4). When λ_S is small, the circumferential velocity is also small which results in an increase in the angle of attack α_A. When the angle of attack increases past a critical angle, the wind flow breaks away from the blade profile and becomes turbulent, thus dramatically reducing the lift force. If the tip speed ratio is too large, the lift force will reach its maximum value and decrease afterwards, thus reducing the power efficiency of the converter.

3.3.3 Power Efficiency

The power efficiency of a rotor blade can be determined by calculating the relationship between the power coefficient and the tip speed ratio. Figure 3.5 shows that for every pitch angle β, there is a tip speed ratio λ_S which corresponds to the maximum power coefficient and hence the maximum efficiency. It can be seen that the power efficiency significantly depends on the pitch angle and the tip speed ratio. Therefore, the pitch angle of the blade has to be changed mechanically in respect to the actual tip speed ratio. In case of any pitch angles the power coefficient is negative. These means that the rotor will not turn. It works as a brake.

A disk brake on the main shaft is therefore not necessary in most modern converters by using pitch control.

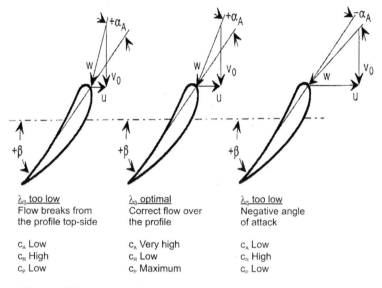

Fig. 3.4   Influence of tip-speed ratio λ_S

Fig. 3.5   Example of the relationship between the power coefficient and the tip-speed ratio

Besides the power coefficient c_p which can be interpreted as the efficiency between the rotor blade and the wind, there are also energy losses in the mechanical components of the rotor and gear system and in the turbine and generator connection. Therefore, the efficiency can be defined as:

$$\eta = c_p \cdot \eta_m \cdot \eta_{ge}, \tag{3.11}$$

with η_m as the mechanical efficiency and η_{ge} as the efficiency of the coupled generator and the electrical auxiliary equipment. The efficiency η is also defined by the relationship of the electrical power to the power potential in the wind:

$$\eta = \frac{P_{el}}{\frac{1}{2} \cdot \rho_a \cdot A \cdot v^3}. \tag{3.12}$$

3.3.4 Principles of Design

The issues discussed above can be summed up and related to the design of a wind energy converter through the following principles:

1. A high aerofoil form ratio c_A/c_R leads to a high tip to speed ratio and therefore a large power coefficient c_p.
   $\longrightarrow$ Modern converters with a good aerodynamic profile rotate quickly.
2. Simple profiles with low aerofoil form ratios have a small tip speed ratio. Therefore, the area of the rotor radius that is occupied by blades must be increased in order to increase the power coefficient.
   $\longrightarrow$ Slow rotating converters have poor aerodynamic profiles and a high number of blades.
3. The profile form ratio and the tip speed ratio have a considerably greater influence on the power coefficient than the number of blades.
   $\longrightarrow$ For modern converters with a good aerodynamic profile, the number of blades is not so important for a large power coefficient c_p.

3.4 Wind Variations

3.4.1 Wind Shear with Height

Assuming that the wind is blowing at 10 m/s at a height of 100 m, Fig. 3.6 shows how wind speeds vary in agricultural land with some houses and sheltering hedgerows with some 500 m intervals.

The fact that the wind profile shown in Fig. 3.6 is twisted towards a lower speed, as one moves closer to ground level, is usually called wind shear. Wind shear may also be important when designing wind turbines. It may be noticed that for a wind

Fig. 3.6 Variation in wind velocity with altitude

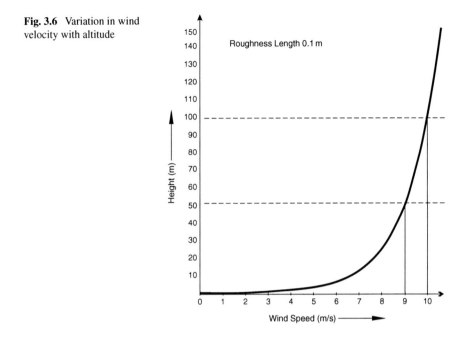

height of 50 m, the wind velocity is 9 m/s whereas for a 100 m height it is 10 m/s in our example case. With the help of the formula for power of wind discussed earlier in this chapter, it can be calculated that due to the dependence of power with the cube of wind speed, this increase of 1 m/s corresponds to about 30% difference in the power available from wind. It is also important to note that if a wind turbine with a hub height of 75 m and a rotor diameter of 50 m is considered, one can notice that the wind is blowing at 10 m/s when the tip of the blade is in its uppermost position (100 m height), and 9 m/s when the tip is in the bottom position (50 m height). This means that the forces acting on the rotor blade when it is in its top position are far larger than when it is in its bottom position.

The wind speed at a certain height above ground level is given by:

$$v = v_{ref} \ln(z/z_0)/\ln(z_{ref}/z_0)$$

where:

v	= wind speed at height z above ground level,
v_{ref}	= reference speed, i.e. a wind speed at height z_{ref},
$\ln(...)$	= natural logarithm function.,
z	= height above ground level for the desired velocity v,
z_{ref}	= reference height, i.e. the height at which wind speed is v_{ref},
z_0	= roughness length in the current wind direction.

Table 3.1 Roughness class and roughness lengths

Roughness class	Roughness length (m)	Energy index (%)	Landscape type
0	0.0002	100	Water surface
0.5	0.0024	73	Completely open terrain with smooth surface, e.g. airport runways, mowed grass
1	0.03	52	Open agriculture land without fencing and very scattered buildings, softly rounded hills
2	0.1	39	Agriculture land with some houses and 8 m tall sheltering hedgerows at a distance of approx. 500 m
3	0.4	24	Villages, small towns,
4	1.6	13	Very large cities with tall buildings and skyscrapers

A List of all major types of roughness classes and their typical roughness lengths is given in Table 3.1.

The above example is assuming that the wind is blowing at 9 m/s at 50 m height and the wind speed at 100 m height is to be calculated. If the roughness length is 0.1 m, then:

$v_{ref} = 9$ m/s
$z \quad = 100$
$z_0 \quad = 0.1$
$z_{ref} = 50$ hence,
$v \quad = 9 \ln(100/0.1)/\ln(50/0.1) = 10$ m/s

Average wind speeds are often available from meteorological observations measured at a height of 10 m. Hub heights of modern 600–1,500 kW wind turbines are usually 40–80 m and more. Using the above mentioned approach, one may calculate average wind speeds at different heights and roughness classes. It is to be noted that the results are not strictly valid if there are obstacles close to the wind turbine (or the point of meteorological measurement) at or above the specified hub height. It should be rather noted that there may be inverse wind shear on hilltops because of the hill effect, i.e. the wind speed may actually decline with increasing height during a certain height interval above the hilltop. A careful study of wind velocity and profile is therefore recommended before arriving at any conclusions about site.

3.4.2 Influence of Weibull Distribution

It is very important to understand the variation of wind speeds within the studied/measured time periods since it governs the energy output significantly. If one measures the wind speeds throughout a year, it can be noticed that in most areas

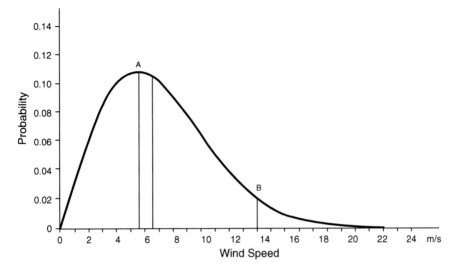

Fig. 3.7   Weibull distribution plot between wind velocity and probability

strong gale-force winds are rare, while moderate and fresh winds are quite common. The impact of such variation on energy output is described in this section.

The wind variation for a typical site is usually described using the so-called Weibull distribution, as shown in Fig. 3.7. This particular site has a mean wind speed of 7 m/s, and the shape of the curve is determined by a so called shape parameter of 2.

It can be realized that the shown graph is a probability distribution. The area under the curve is always exactly 1, since the probability that the wind will be blowing at some wind speed must be 100%.

Half of the area is to the left of the vertical black line at 6.6 m/s, and half is on its right hand side. The 6.6 m/s is called the median of the distribution shown above. From site-to-site, this median value would be differing. For this site, it means that half the time it will be blowing less than 6.6 m/s, the other half it will be blowing faster than 6.6 m/s. As it can be seen, the distribution of wind speed is skewed, i.e. it is not symmetrical. Sometimes there will be very high wind speeds, but they are very rare. Wind speeds of 5.5 m/s, on the other hand, are the most common ones. The statistical distribution of wind speed varies from place to place around the globe, depending upon local climate conditions, the landscape, and its surface.

The Weibull distribution may vary from site to site, both in its shape, and in its median value. There are two parameters that govern the shape of the Weibull distribution curve, namely, the scale parameter and the shape parameter. A higher value of the scale parameter means the distribution is spread over a wider range and the probabilistic average wind velocity has a higher value. A higher value of shape parameter (between 2 and 3) means the distribution is more skewed towards higher wind velocities, if the shape parameter is between 1 and 2, it means that the distribution is skewed towards lower velocities, indicating a higher probability of lower wind velocities. Of course, both these parameters influence the peak distribution curve, but one has major influence on the average value and the other primarily influences the skewness of the curve. For an exact feeling of curve, both should be considered together.

If the shape parameter is exactly 2, the distribution is known as a Rayleigh distribution. Wind turbine manufacturers often give standard performance figures for their machines using the Rayleigh distribution.

The reason why care is to be taken in connection with wind speeds is their energy content. The wind power varies with the cube of the wind speed. Design/selection of a wind turbine as per mathematical average wind speed may lead to an improper utilization of wind potential at the site. In Fig. 3.7, two conditions marked A and B, one corresponding to the peak velocity i.e. 5.5 m/s which has a probability of 11%, and the other the velocity of 14 m/s having a probability of 2% can be compared. Due to the dependence of power with cube of velocity, at these two velocities 5.5 m/s and 14 m/s, the difference between their cubes, i.e. between 166 and 2,744, is huge. Which means that at the high speed of 14 m/s, more than 16 times the power is available but only for 2% duration, whereas for a large duration of 11% only a small amount of power is available. If the wind turbine is selected as per 14 m/s wind velocity, it would remain much underutilized since the probability of the presence of smaller wind velocities is much higher. However, these figures are to be seen with respect to the energy delivered by the wind turbine and the time/day/month of delivery. Some countries such as Germany, where a feed-in tariff system exists, only the amount of power delivered is important, whereas, at some places the time of delivery may also be important. This analysis that affects the selection of a machine very much, can be done through the power curve of the machine being selected, which is discussed in Chap. 6.

Chapter 4
Components of a Wind Energy Converter

A wind energy converter has the following major components:

1. Rotor blades
2. Gearbox
3. Generator
4. Tower
5. Miscellaneous parts

All of these parts, their major design features and variations available on the market are discussed in this chapter. Besides these major components, there are other necessary parts, such as grid connection. Options for grid connection are covered in Chap. 6.

4.1 Rotor Blades

The basic principles behind the working of wings of airplanes and blades of wind turbines are quite common. However, since the wind turbines actually work in a very different environment with changing wind speeds and changing wind directions, there are special considerations that are not important in the design of airplane wings.

Figure 4.1 shows a photo of a rotor blade for a 1.5 MW wind converter. Figure 4.2 defines the four sides of a blade.

Due to the aerodynamic profile of blades, as discussed in Chap. 3, low pressure is created on the upper surface of a blade. This creates the lift, i.e. the force pulling upwards, the same principle that enables the plane to stay in the air. In case of the rotor blade of a wind mill, the lift is perpendicular to the direction of the wind.

Choosing profiles for rotor blades involves a number of compromises including reliable pitch control and stall characteristics, and the profile's ability to perform well even if there is some dirt on the surface. In countries where rains occur during the whole year, the dirt gets cleaned by the rain water, but in relatively dry countries like India, dirt may become a problem. Even in Germany, when there is only light rain, it may not be sufficient to whip-off the dirt from the blades. Under these

H.-J. Wagner, J. Mathur, *Introduction to Wind Energy Systems*, Green Energy and   29
Technology, DOI 10.1007/978-3-642-02023-0_4, © Springer-Verlag Berlin Heidelberg 2009

Fig. 4.1 Rotor blade of a 1.5 MW wind turbine (photo: Wagner)

Fig. 4.2 Four sides of a rotor
blade of a wind turbine

circumstances, the blades will have to be cleaned and possibly even polished again
after 4–6 years, to ensure a smooth aerodynamic effect.

Most modern rotor blades on large wind turbines are made of glass fibre reinforced plastics, (GRP), i.e. glass fibre reinforced polyester or epoxy. Using carbon

fibre or aramid (Kevlar) as reinforcing material is another possibility, usually such blades become necessary in large turbines in the range of 3–6 MW capacity. Wood, wood-epoxy, or wood-fibre-epoxy composites have not penetrated the market for rotor blades. Steel and aluminum alloys have problems of weight and metal fatigue respectively.

Large rotor blades typically consist of a metal walled circular root section with metal bushings with bolts or studs for mounting the blade to the hub. This root section is integrated into a continuously tapered longitudinal beam, the spar, which provides the stiffness and strength required to carry the wind load and the weight of the blade. Around the spar, the two aerodynamic-shaped shells, the suction side and the pressure side, form an optimized aerodynamic wing. The outer shells of the blade meet at the leading edge and the trailing edge. Inserted webs take up the torsional twist of the blade and help stabilize the blade against bending, shear loads and global buckling.

4.2 Gearboxes

The power from the rotation of the wind turbine rotor is transferred to the generator through the power train, i.e. through the main shaft, the gearbox and the high speed shaft. It is not appropriate to just drive the generator directly with the power from the main shaft unless the turbine is equipped with multi-pole generators that are discussed in the next chapter.

If an ordinary generator with two poles (which makes one pair of poles) was used without a gearbox, and the machine was directly connected to a 50 Hz AC (alternating current) three-phase grid, it would become necessary to turn the rotor at 50 revolutions per second i.e. 3,000 revolutions per minute (rpm). With a 50 m rotor diameter that would imply a tip speed of the rotor of far more than twice the speed of sound, a condition which cannot be accepted. Usually, a tip-speed of more than 100 m/s is not accepted across the industry. Another possibility of avoiding a gearbox is to build a slow-moving AC generator with a lot of poles. If a generator was to be connected directly to the grid, this would require a 300 pole generator to arrive at a reasonable rotational speed of 20 rpm.

With a gearbox, slowly rotating, high torque power obtained from the wind turbine rotor can be converted into high speed power, which would be required for the generator. The gearbox in a wind turbine usually does not "change gears". It normally has a single gear ratio between the rotation of the rotor and the generator. In some machines, the gear ratio is in the range of 30–200.

A typical issue for a gearbox is cooling and heating of the lubricant. In cold climatic conditions like in Germany, in winters, the lubricating oil freezes due to extreme low temperatures. Special heating arrangements are made for keeping the oil sufficiently warm. The situation in hot climates, like in India, is reverse. Arrangements are to be made to prevent the lubricating oil from becoming hot.

A recent development in the field of gearboxes for modern machines is the development of hydrodynamic gearboxes that are actually a combination of a conventional mechanical gearbox and a torque changer. There is a variable speed drive inside and a constant drive outside. The main advantage of this type is continuously variable input but constant output speed adjustment over a wide range. Prototypes have been successfully tested and this technology is very close to the commercialization stage.

4.3 Generators

The wind turbine generator converts mechanical energy to electrical energy. Wind turbine generators are a bit different, compared to other generating units ordinarily attached to the electrical grid. One reason is that the generator has to work with a power source (the wind turbine rotor) which supplies very fluctuating mechanical power (torque).

4.3.1 Generating Voltage

In large wind turbines the voltage generated by the turbine is usually around 690 V three-phase alternating current (AC). The current is subsequently sent through a transformer next to the wind turbine (or inside the tower) to raise the voltage to somewhere between 10,000 and 30,000 V, depending on the standard in the local electrical grid. Manufacturers will supply both 50 Hz wind turbine models (for the electrical grids in most of the world) and 60 Hz models (for the electrical grid e.g. in America).

4.3.2 Generator Cooling System

Generators need cooling while they work. On most turbines this is accomplished by encapsulating the generator in a duct, using a large fan for cooling by air, but a few manufacturers use hydraulically cooled generators. Hydraulically cooled generators may be built more compactly, but they require a heat exchanger (radiator) in the nacelle to get rid of the heat from the liquid cooling system.

4.3.3 Generator Rotational Speed

The speed of a generator which is directly connected to a grid is constant, and dictated by the frequency of the grid. The relationship between the rotational speed of generator and frequency is governed by the formula:

$$\text{Rotation per second} = \frac{\text{Frequency per second (50 Hz or 60 Hz)}}{2 \times \text{ number of poles}}. \qquad (4.1)$$

Table 4.1 Synchronous Generator Speeds in records per minute (rpm)

Pole number	No. of pole pairs	50 Hz	60 Hz
2	1	3,000	3,600
4	2	1,500	1,800
6	3	1,000	1,200
8	4	750	900
10	5	600	720
12	6	500	600
300	150	20	–
360	180	–	20

The number of poles means the number of coil sets in the stator of the generator which the electric power is generated in. Two poles make one pole pair. The number of poles will be decided by the construction of generator. Because of the fact that grid is operating with three-phases, the generator must also produce three phase alternating current.

When the number of pole pairs is doubled in the stator of a synchronous generator, the rotation will be reduced to half. The variations between rotation per minute and pole pairs are given in Table 4.1.

Often wind turbines that are with gearboxes, use generators with two or three pole pairs. The reasons for using these relatively high-speed generators are savings on size and cost. The maximum force (torque) a generator can handle depends on the rotor volume. The other type of wind turbines that do not use a gearbox but use multi-pole generators (explained in the next chapter), have a large number of poles.

The number of poles is kept less than 300 (say 84) due to weight considerations. As a result, the frequency of electricity generated is less than 50 Hz, which is transformed into 50 Hz before supply to the grid electronically.

The term "synchronous generator speed " (f/p, the ratio of frequency and number of pole pairs) thus refers to the speed of the generator when it is running synchronously with the grid frequency. It applies to synchronous generators . However, in the case of asynchronous (induction) generators   it is equivalent to the idle speed of the generator. This means that the generator rotation must be a little bit higher than the synchronous speed e.g. in case of 2 pole pairs, 1,515 rpm for a synchronous speed of 1,500 rpm. The advantages and disadvantages of synchronous and asynchronous generators are further discussed in Chap. 5.

Wind turbines which use synchronous generators use electromagnets normally in the rotor which are excited by direct current from the electrical grid. Since the grid supplies alternating current, they first have to convert alternating current to direct current before sending it into the coil windings around the electromagnets in the rotor. The rotor electromagnets are connected to the current by using brushes and slip rings on the axle (shaft) of the generator.

Permanent magnet synchronous generators are tested in prototypes but not used so often despite their advantage of having a low weight. There are several reasons for this. One reason is the fear that permanent magnets could become demagnetized because of the mechanical forces. Another reason is that powerful magnets (made of rare earth metals, e.g. Neodymium) are quite expensive, although prices have dropped lately.

The key component of the basic version of an asynchronous generator is the squirrel cage rotor or simple cage rotor.

For understanding the operation of an asynchronous generator in basic version, consider manual cranking of this rotor around at exactly the synchronous speed of the generator, e.g. 1,500 rpm, for the 2-pole pair synchronous generator. Since the magnetic field rotates at exactly the same speed as the rotor, there would be no induction phenomena in the rotor, and it would not interact with the stator. The stator is connected to the grid and its field is rotating recording the grid frequency and pole numbers with 1,500 rpm. If the speed of rotor is increased above 1,500 rpm, in that case the rotor moves faster than the rotating magnetic field from the stator, in that condition, once again the stator induces a strong current in the rotor. The harder the rotor is cranked, the more power will be transferred as an electromagnetic force to the stator. The current in the rotor will create an electromagnetic field which generates electricity in the stator, which is fed into the electrical grid.

It is the rotor that makes the asynchronous generator different from the synchronous generator. The rotor consists of a number of copper or aluminum bars which are connected electrically by copper or aluminum end rings. The rotor is placed in the middle of the stator which is directly connected to the three-phases of the electrical grid. The clever thing about the cage rotor is that it adapts itself to the number of poles in the stator automatically. The same rotor can, therefore, be used with a wide variety of pole numbers.

4.3.4 Two Speed, Pole Changing Generators

Some manufacturers fit their turbines with two generators, a small one for periods of low winds, and a large one for periods of high winds. A more common design on newer machines is pole changing generators, i.e. generators which (depending on how their stator windings are connected) may run with a different number of pole pairs, and thus a different rotational speed.

Whether it is worthwhile to use a switching of pole pairs in the generator depends on the local wind speed distribution, and the extra cost of the pole pairs changing generator compared to the price the turbine operator gets for the electricity. However, due to the problems related to a drop in grid voltage, the practice of switching the poles has been discontinued in Europe.

4.4 Towers

The tower of the wind turbine carries the nacelle and the rotor blades. The tower needs to be as tall as possible, because the wind speed increases with height. However, the height is optimized by analyzing the cost of the increase in tower height and the gain in energy output due to increased wind velocity at a greater height. Towers for large wind turbines may be either tubular steel towers, lattice towers, or concrete towers. Guyed pole towers are only used for small wind turbines.

(a) *Tubular steel towers.* Most large wind turbines are delivered with tubular steel towers, which are manufactured in sections of 20–30 m with flanges at either end, and bolted together on the site. The towers are conical (i.e. with their diameter increasing towards the base) in order to increase their strength and to save materials at the same time.

(b) *Lattice towers.* Lattice towers are manufactured using welded steel profiles. The basic advantage of lattice towers is cost, since a lattice tower requires only half as much material as a freely standing tubular steel tower with a similar stiffness.

The advantage of the low cost of lattice towers exists only in those countries where the labor cost is cheap e.g. in India. In countries like Germany, where the labor costs are high, the advantage of reduced material cost gets balanced by increased labor costs for assembling the tower. Another disadvantage of lattice towers is their visual appearance, although that issue is clearly debatable. Be that as it may, for aesthetic reasons lattice towers have almost disappeared from being used for large, modern wind turbines. These towers are used only in small machines i.e. below the MW size.

(c) *Concrete towers.* Recently, with a further increase in the height of towers for more than 100 m hub height, due to the increased cost of steel in towers, manufacturers have started to bring concrete towers in place of steel towers on the market. In future, therefore, there may be an additional type of tower on the wind energy market.

(d) *Guyed pole towers.* Many small wind turbines are built with narrow pole towers supported by guy wires. The advantage lies in saving weight and, consequently, cost. The disadvantage lies in a difficult access around the towers which makes them less suitable in farm areas. Such towers are used only in kW size machines.

(e) *Hybrid tower solutions.* Some towers are built by using different combinations of the techniques mentioned above. It may be a hybrid between a concrete tower (down) and steel (upward) or between a lattice tower and a guyed tower.

Generally, it is an advantage to have a tall tower in areas with a high terrain roughness, since the wind speeds increase with height above the ground, as discussed in previous chapters. Lattice towers and guyed pole towers, due to a reduced surface area as compared to tubular steel towers, have the advantage of giving less wind shade than a massive tower. However, the turbulence created by each element of a lattice structure adversely affects the energy output.

4.5 Miscellaneous Components

Large wind turbines are equipped with a number of safety devices to ensure safe operation during their lifetime. A brief description of such devices is given below:

(a) *Yaw mechanism.* Almost all horizontal axis wind turbines use forced yawing, i.e. they use a mechanism which uses electric motors and gearboxes to keep the rotor plane perpendicular to the direction of the wind. Almost all manufacturers of upwind machines prefer to apply brakes to the yaw mechanism whenever it is unused. The yaw mechanism is activated by the electronic controller which checks the position of the wind vane on the turbine several times per minute, whenever the turbine is running.

Cables carry the current from the wind turbine generator down through the tower. Occasionally, a wind turbine may look as if it had gone berserk, yawing continuously in one direction for about any revolutions. The cables, however, will become more and more twisted if the turbine by accident keeps yawing in the same direction for a long time. The wind turbine is, therefore, equipped with a cable twist counter, which tells the controller that it is time to untwist the cables. The turbine is also equipped with a pull switch which gets activated if the cables become too twisted.

Besides the role of tracking the wind direction, the yaw mechanism plays another very important role by connecting the tower with the nacelle. Due to the operation of gears, it needs regular lubrication for smooth operation.

(b) *Brakes*. Braking action may be required for several reasons. There are several types of braking a turbine rotor: aerodynamic brakes , electro brakes and mechanical brakes. In case of aerodynamic braking, the blade is turned in such a direction that the lift effect which causes rotation, does not appear. This concept is explained in the section of "stall control" in Chap. 6. In case of electro-magnetic braking, energy produced by the generator of a wind turbine is dumped into a resistor bank, thereby converting it into heat.

Another type of braking is conventional mechanical braking for which disc brakes are provided in the nacelle. A mechanical drum brake or disk brake is also used to hold the turbine at rest for maintenance. Such brakes are also applied after blade furling and electro braking have reduced the turbine speed, as the mechanical brakes would wear quickly if used to stop the turbine from full speed.

In large wind turbines, normally there is a combination of at least two brakes, most turbines use aerodynamic brakes together with mechanical braking or even also with an electro braking system.

(c) *Over-speed protection*. In addition to braking explained above, it is essential that wind turbines stop automatically in case of malfunction of a critical component, for example, if the generator overheats or is disconnected from the electrical grid and the rotor will start accelerating rapidly within a matter of seconds. In such a case it is essential to have an over-speed protection system. Most companies provide wind turbines with two independent fail-safe brake mechanisms to stop the turbine.

(d) *Vibration measurement*. Excessive vibrations may be one of the most dangerous causes for failure of a wind turbine as well as any other component. Vibrations cause fatigue loading of components causing early failure. An additional side effect of vibration is the generation of noise. One of the classical, and most simple safety devices in a wind turbine is the vibration sensor. It simply consists of a ball resting on a ring. The ball is connected to a switch through a chain. If the turbine starts shaking, the ball will fall off the ring and switch the turbine off. Modern turbines have permanently installed vibration monitoring systems which in the case of excessive vibrations decide on the operation of the machine with the help of microprocessors.

(e) *Anemometer.* Before actuating the controlling actions, measurement of wind speed and direction is necessary in every wind turbine. It is usually done using a cup anemometer . The cup anemometer has a vertical axis and three cups which capture the wind. The number of revolutions per minute is registered electronically. Normally, the anemometer is fitted with a wind vane to detect the wind direction. Other anemometer types include ultrasonic or laser anemometers which detect the phase shifting of sound or coherent light reflected from the air molecules. Hot wire anemometers detect the wind speed through minute temperature differences between wires placed in the wind and in the wind shade (the leeward side). The advantage of non-mechanical anemometers may be that they are less sensitive to icing in cold countries and jamming due to dust in hot locations.

(f) *Lubrication system.* Since the wind turbine has moving parts, lubrication is always a major concern. There are two types of lubrication systems . With the first type, there is a central lubrication pump which sends lubricant to all bearings through tubes laid for this purpose. The other system has a pressurized container of lubricant for each bearing which directly lubricates the bearings.

(g) *Foundation.* Foundation plays a very important role in stabilizing the wind turbine. Due to the large height, heavy weight at the nacelle and large rotor area which faces wind forces, the role of a foundation becomes very important. While in case of on-shore installations, the type of foundation depends upon the nature of the soil, in case of off-shore turbines, it becomes an even more serious issue. Its designs tends to rely on technology used by the oil and gas industry in some cases. There are different major types of foundations used in case of off-shore wind turbines. The first is the mono pile foundation (see Fig. 4.3) that consists of a steel pile which is

Fig. 4.3 Three foundations for off-shore wind turbines. From left to right: Monopile, Tripod, Jacked

Fig. 4.4 Tripod foundation for the windpark Alpha Ventus in Germany (photo: Große Böckmann)

driven approximately 10–12 m into the seabed. The second one is the Tripod foundation (Fig. 4.4). The piles on each end are typically driven about 10 m into the seabed, depending on soil conditions. The third is a so called jacked . It is having a lattice tower also, fixed by piles driven in the seabed. The fourth is a so called tripile solution (Fig. 4.5). With the tripile solution the advantages of the simpler monopole construction and transportation can be used, but the necessary load distribution on tops must designed carefully, concerning consequences of possible motions. The next one is the gravity foundation . It consists of a large base constructed from concrete which rests on the seabed. The turbine is dependent on gravity to remain erect.

Under development is another type of foundation, a swimming platform on which the wind tower is installed. The platform will swim a little bit under the water surface and will be fixed by guy wires on sea ground.

Monopiles and gravity foundation seems to be used in water depth less than about 20 m. The other foundation constructions will also be used in deeper water.

(h) *Other parts*. There are many other sensors in the nacelle, e.g. electronic thermometers which check the oil temperature in the gearbox and the temperature of the generator. There are other important measurements that play a major role in connection and operation of a wind turbine with grid: voltage-frequency measurement, phase-angle measurement, revolution measurement, electric power supply measurement. There are self controlling systems with a telecommunication link to a

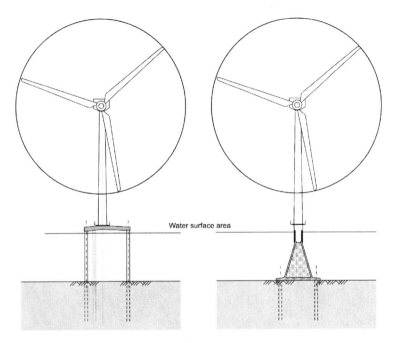

Fig. 4.5   Tripile and gravity foundation for off-shore wind turbines

common service point. All of these make the wind turbine more efficient and available for a maximum time for the economic generation of power. The components related to grid connection are discussed in Chap. 6.

Chapter 5
Design Considerations

When finalizing the specifications of wind turbines for installation at any site, a lot of technological options are found to be available. These options along with a discussion on their suitability in various situations are presented in this chapter.

5.1 Rotor Area of Turbines

The question of small or large turbines pertains to the decision related to the area of rotor. A larger rotor with compatible generator would be called a large turbine.

5.1.1 Reasons for Choosing Large Turbines

1. There are economies of scale in wind turbines, i.e. larger machines are usually able to deliver electricity at a lower cost than smaller machines. The reason is that the cost of foundations, road building, electrical grid connection, plus a number of components in the turbine (the electronic control system etc.) are somewhat independent of the size of the machine.
2. Larger machines are particularly well suited for offshore wind power. The cost of foundations does not rise in proportion to the size of the machine.

5.1.2 Reasons for Choosing Smaller Turbines

1. The local electrical grid may be too weak to handle the electricity output from a large machine. This may be the case in remote parts of the electrical grid with a low population density and little electricity consumption in the area.
2. The cost of using large cranes, and building a road strong enough to carry the turbine components may make smaller machines more economic in some areas.
3. Aesthetic landscape considerations are dependent upon the perception of local people. Some people like to see larger machines and attach them with their own identity. Another set of people may not want to see any extraordinary large structures in the natural landscape. Speed of rotation also plays an important role

H.-J. Wagner, J. Mathur, *Introduction to Wind Energy Systems*, Green Energy and Technology, DOI 10.1007/978-3-642-02023-0_5, © Springer-Verlag Berlin Heidelberg 2009   41 in perception about landscape, since large machines have a much lower rotational speed, which means that one large machine really does not attract as much attention as many small, fast moving rotors.

In countries like Germany, where the availability of a strong grid to accept power is not a problem, the question of a small or large rotor area becomes insignificant. There is a clear cut tendency to prefer larger machines due to the economic advantage with a large rotor diameter. However, the selection between large and small machines becomes an important criterion in developing countries like India where the selection of larger machines may become inappropriate in the absence of availability of a strong grid.

5.2 Number of Blades

In modern wind turbines, use of an even number of rotor blades is avoided. The most important reason is the stability of the turbine. A rotor with an odd number of rotor blades (and at least three blades) can be considered to be similar to a disc when calculating the dynamic properties of the machine. A rotor with an even number of blades will give stability problems for a machine with a stiff structure. The reason is that at the very moment when wind is coming from the front, the uppermost blade bends backwards, because it gets the maximum power from the wind. If an even number of blades is used, at that time the lowermost blade passes into the wind shade in front of the tower. This combination makes the dynamic balancing of rotor more well-balanced.

Most modern wind turbines, therefore, are three-bladed designs. The vast majority of the turbines sold in world markets have this design. Selection of the number of blades depends upon the wind profile.

Besides the above mentioned disadvantage, two-bladed wind turbine designs have the advantage of saving the cost of one rotor blade and its weight, of course. However, they tend to have difficulties in penetrating the market, partly because they require higher rotational speed to yield the same energy output. This is a disadvantage both in regard to noise and visual intrusion. Lately, several traditional manufacturers of two-bladed machines have switched to three-bladed designs.

One-bladed wind turbines also exist as prototypes, and indeed, they save the cost of another rotor blade. One-bladed wind turbines are not very widespread commercially, however, because the same problems that are mentioned with the two-bladed design apply to an even larger extent to one-bladed machines. In addition to the higher rotational speed, and the noise and visual intrusion problems, they require a counterweight to be placed on the other side of the hub from the rotor blade in order to balance the rotor. This obviously negates the savings on weight compared to a two-bladed design.

Two- and one-bladed machines require a more complex design with a hinged (teetering hub) rotor. The rotor has to be able to tilt in order to avoid too heavy

Fig. 5.1 Power coefficients and tip-speed ratio of different types of rotors (A: Savonius rotor, B: Multi blade rotor, C: Four blade rotor, D: Three blade rotor, E: Two blade rotor)

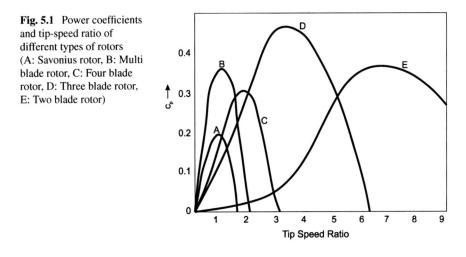

shocks to the turbine when a rotor blade passes the tower. The rotor is, therefore, fitted onto a shaft which is perpendicular to the main shaft, and which rotates along with the main shaft. This arrangement may require additional shock absorbers to prevent the rotor blade from hitting the tower.

Due to mechanical problems, two blade and single blade designs have not become very popular. Additionally, it is understood that the three blade design is psychologically more acceptable to human perception as compared to the two or single blade design. Figure 5.1 shows the range of power coefficient and tip speed ratio for different types of rotors. It shows that as compared to the three blade rotors, the two blade rotors need a higher tip-speed ratio for operating at the same value of power coefficient. This means that for a given wind velocity, a two blade rotor has to rotate faster as compared to a three blade rotor.

5.3 Horizontal or Vertical Axis Turbine

Considering the present status of trade, the choice between horizontal axis and vertical axis wind turbines is almost theoretical and goes in favor of horizontal axis turbines. However research is going on to bring back vertical axis turbines into practice, which may be possible in future.

(a) *Horizontal axis wind turbines.* Most of the technology described on these pages is related to horizontal axis wind turbines. The reason is simple: All grid-connected commercial wind turbines today are built with a propeller-type rotor on a horizontal axis (i.e. a horizontal main shaft). The purpose of the rotor, of course, is to convert the linear motion of the wind into rotational energy that can be used to drive a generator. The same basic principle is used in a modern water turbine, where the flow of water is parallel to the rotational axis of the turbine blades.

Fig. 5.2 A 4.2 MW vertical
axis Darrieus wind turbine
(*source:* www.reuk.co.uk)

(b) *Vertical axis wind turbines.* Figure 5.2 shows the biggest prototype of a vertical axis wind turbine, with a 100 m rotor diameter and a capacity of 4.2 MW at Cap Chat, Québec, Canada. The machine is no longer operational. It was operational in the period 1983–1992. The only vertical axis turbine which has ever been manufactured commercially at any volume is the Darrieus machine, named after the French engineer Georges Darrieus who patented the design in 1931. It was manufactured by the U.S. company FloWind which stopped its production in 1997. The Darrieus machine is characterized by its C-shaped rotor blades. It is normally built with two or three blades. Another type of vertical axis machine is the machine with H-rotor. The name is derived from the shape "H" of its rotor (see Fig. 5.3). One more type of vertical axis machine consists of Savonius rotors which are primarily used for measuring weather conditions.

The basic theoretical advantages of a vertical axis machine are:

1. The generator, gearbox etc. can be placed on the ground, and may not need a tower for the machine.
2. A yaw mechanism is not needed to turn the rotor against the wind.

The basic disadvantages are:

1. Wind speeds are very low close to ground level, so the need of a tower is eliminated, but the wind speeds will be very low on the lower part of the rotor.

Fig. 5.3 H-rotor in Germany
(photo: Große Böckmann)

2. The overall efficiency of the vertical axis machines is less than in the case of a horizontal axis.
3. The machine is not self-starting, e.g. a Darrieus machine will need a "push" before it starts. This is only a minor inconvenience for a grid connected turbine, since the generator may be used as a motor drawing current from the grid to start the machine.
4. The machine may need guy wires to hold it up, but guy wires are impractical in heavily farmed areas.

5.4 Upwind or Downwind Turbine

Upwind machines are those machines that have the rotor facing the wind. In these machines the wind meets the rotor first and then leaves from the direction in which the nacelle is located. Downwind machines have the rotor placed on the leeward side of the tower; this means the nacelle comes first in the path of the wind and then the blades, as shown in Fig. 5.4.

(a) *Upwind machines*. The basic advantage of upwind designs is that one avoids the wind shade behind the tower. By far the vast majority of wind turbines have this design. On the other hand, there is also some wind shade in front of the tower,

Fig. 5.4 Upwind and
downwind type wind turbines

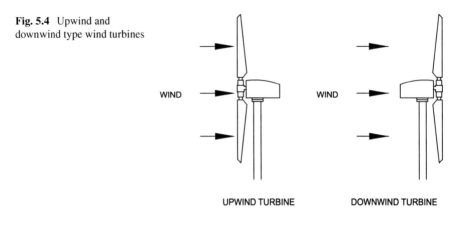

i.e. the wind starts bending away from the tower before it reaches the tower itself,
even if the tower is round and smooth. Therefore, each time the rotor passes the
tower, the power from the wind turbine drops slightly. The basic drawback of
upwind designs is that the rotor needs to be placed at some distance from the tower.
In addition, an upwind machine needs a yaw mechanism to keep the rotor facing
the wind.

(b) *Downwind machines*. They have the theoretical advantage that they may be
built without a yaw mechanism, if the rotor and nacelle have a suitable design
that makes the nacelle follow the wind passively. For large wind turbines this is
a somewhat doubtful advantage, since for optimal energy efficiency of wind energy
converters, the yaw control must be applied very accurately.

Another advantage of the downwind design is that the rotor may be made more
flexible. This is an advantage in regard to weight and the structural dynamics of the
machine, i.e. the blades will bend at high wind speeds, thus taking part of the load
off the tower.

The basic drawback is the fluctuation in the wind power due to the rotor passing
through the wind shade of the tower. This may give more fatigue loads on the turbine
than with an upwind design.

Out of these two options, upwind machines are more commonly used due to an
increased energy output and hence their being more economical.

5.5 Load Considerations for Turbine Selection

Wind turbines are subject to fluctuating winds, and hence fluctuating forces. This is
particularly the case if they are located in a very turbulent wind climate. Components
which are subject to repeated bending, such as rotor blades, may eventually develop
cracks which ultimately may make the component break. Metal fatigue is a well
known problem in many technical goods. Due to combined reasons of fatigue and
mass, metal is, therefore, generally not preferred as a material for rotor blades. When

designing a wind turbine it is extremely important to calculate in advance how the different components will vibrate, both individually, and jointly. It is also important to calculate the forces involved in each bending or stretching of a component. This is the subject of structural dynamics, where mathematical computer models have been developed that analyse the behavior of an entire wind turbine. These models are used to design the machines safely.

As an example, a tall wind turbine tower will have a tendency to swing back and forth, say, every three seconds. The frequency with which the tower oscillates back and forth is also known as the eigenfrequency of the tower. The eigenfrequency depends on the height of the tower, the thickness of its walls, the type of steel, and the weight of the nacelle and rotor. Now, each time a rotor blade passes the wind shade of the tower, the rotor will push slightly less against the tower.

If the rotor turns with a rotational speed such that a rotor blade passes the tower each time the tower is in one of its extreme positions, then the rotor blade may either dampen or amplify (reinforce) the oscillations of the tower. The rotor blades themselves are also flexible, and may have a tendency to vibrate, say, once per second. Consequently, it is very important to know the eigenfrequencies of each component in order to design a safe turbine that does not oscillate out of control.

5.6 Wind Turbines: With or Without Gearbox

(a) *Design with gearbox.* The principle of a design of a wind turbine with gearbox, is shown in Fig. 5.5 and 5.6. The main aspect of this design is the split shaft system, where the main shaft turns slowly with the rotor blades and the torque is transmitted through a gearbox to the high-speed secondary shaft that drives the few-pole pair generator. The transmission of torque to the generator is shut off by means of a large disk brake on the main shaft. A mechanical system controls the pitch of the blades, so pitch control can also be used to stop the operation of the converter, e.g. in stormy conditions. The pitch mechanism is driven by a hydraulic system, with oil as the popular medium. For constructions without a main brake, each blade has its pitch angle controlled by a small electric motor.

The gearbox concept was in many cases accompanied by an insuffcient life time because of failure of gearboxes. After many years of operational experiences and a lot of research and development activities it got solved.

(b) *Design without gearbox.* Some companies, e.g. the German company Enercon, designed another converter type without gearbox. The scheme of such a converter is shown in Fig. 5.7, where the main design aspects can be clearly seen. This design has just one stationary shaft. The rotor blades and the generator are both mounted on this shaft. The multi-pole generator is in the form of a large spoked wheel with e.g. forty-two pole pairs around the outer circumference and stators mounted on a stationary arm around the wheel. The wheel is fixed to the blade apparatus, so it rotates slowly with the blades. Therefore, there is no need for

Fig. 5.5 Wind turbine with a gearbox

a gearbox, rotating shafts or a disk brake. This minimizing of rotating parts reduces maintenance and failure possibilities and simplifies the maintenance and production of the converter. The price for this advantages is a high nacelle mass caused by the high copper content of multi-pole generator.

(c) *Multibrid design*. Both designs explained above have one disadvantage each. The design with gearbox has the disadvantage of losses during transmission of power and high speed drive required for connecting the generator, while the gearless design has the disadvantage of increased weight of the nacelle due to an increased number of poles. The objective of the multibrid design is to overcome disadvantages of both previous designs. In this design, a new combination of approved techniques has been used. It is operated in the M5000 design of the research and development company Multibrid (a registered name) and was developed especially for offshore use. This design is a combination between a special gearbox and a multi-pole gener-ator. There is no high-speed transmission included in the gearbox, which is critical to failure. The selection of transmission achieves a high transmission allowing the

Fig. 5.6 Assembling of a wind turbine with gearbox by NORDEX company (photo: Nordex)

Fig. 5.7 Wind turbine without gearbox (Design of ENERCON company)

use of a generator with up to 150 rpm. Additionally, in the prototypes, a permanent magnet-excited synchronous generator with water-cooling, high efficiency and wide speed range is used. Due to the compact design, the construction has a reduced weight.

5.7 Requirement of Grid, Synchronous or Asynchronous Generators

The main electrical grid has a constant frequency, e.g. 50 Hz, and a constant phase angle. Therefore, a wind energy converter must produce electricity with the same constant values in order to integrate with electricity in the main grid. The input energy of a wind energy converter is proportional to the wind speed, but the wind speed is never constant. Each wind speed has a corresponding rotor rotation speed at which the maximum power is produced, as shown in Fig. 5.8. This maximum occurs at different rates of rotation for different wind speeds. However, the rate of rotation must be kept constant in order to achieve the required constant output frequency, e.g. by 18.5 rpm in the figure. Solutions to this problem of maximizing the power output of converters by variable speeds and constant frequency are discussed in connection with the generator.

With respect to their working principle, there are two choices for generators: synchronous generators or asynchronous generators.

(a) *Synchronous generators*. All generators use a rotating magnetic field for the generation of electricity. The more force (torque) is applied and the rotational speed is kept constant, the more electricity is generated, but the synchronous generator will still run at the same speed dictated by the frequency of the electrical grid.

If the generator is disconnected from the main grid, however, it will have to be cranked at a constant rotational speed in order to produce alternating current with a constant frequency. The synchronous generator is generating power with constant frequency. Consequently, with this type of generator it is normally recommended to use an indirect grid connection for being able to operate it with different rotational speeds.

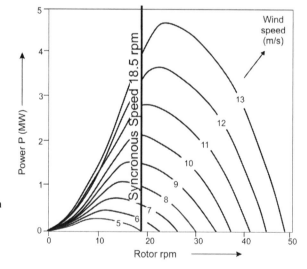

Fig. 5.8 Change in optimum rotation with wind speed for maximum power (synchronous speed 18.5 rpm)

Fig. 5.9 Grid connection of
synchronous generators (SG)
with full inverter system

a) Grid connection for synchronous generators with gearbox

n= (0.5 to 1.2) (f/p) controllable
Inductive reactive load
Controllable reactive power

b) Grid connection for synchronous generators without gearbox

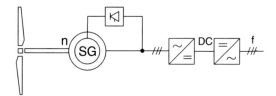

n= (0.5 to 1.2) (f/p) controllable
Inductive reactive load
Controllable reactive power

Figure 5.9 shows the connection of a wind turbine with synchronous generator
to the grid. The synchronous generator operates with a variable speed of rotation,
which means that it will produce electric power with variable frequencies. Due to a
DC link, this will be transformed to AC power with the required frequency (usually
the grid frequency). Use of DC has no link with the use of design having gearbox or
without gearbox as shown in Fig. 5.9a, b for both cases.

(b) *Asynchronous (induction) generators.* The other option for choosing a generator is a three phase asynchronous generator, also called an induction generator
to generate alternating current. This type of generator is widely used in the wind
turbine industry and in small hydropower units. The unique feature of this type of
generator is that it was really originally designed as an electric motor. One reason
for choosing this type of generator is that it is very reliable, and tends to be com-
paratively inexpensive with the basic design of a cage rotor. The generator also has
some mechanical properties which are useful for wind turbines such as a generator
slip, and a certain overload capability.

The speed of the asynchronous generator must be little higher than the synchronous speed. In practice, the difference between the rotational speed at peak
power and at idle is very small, up to 8%. This difference in per cent of the synchronous speed is called the generator's slip. Thus, a 4-pole generator will run idle
at 1,500 rpm if it is attached to a grid with a 50 Hz current. If the generator is producing at its maximum power, lets say with 1% slip, it will be running at 1,515 rpm.

Fig. 5.10 Grid connection of   a) Direct grid connection
asynchronous generators
(ASG)

n = (1+s) (f/p), s = 0 to 0.08 (Power dependent)
Inductive reactive load

b) Dynamic slip control

n = (1+s) (f/p), s = 0 to 0.1
Inductive reactive load

c) Double excited asynchronous generator

n = (1+s) (f/p), s = 0 to 0.3
Controllable reactive power

It is a very useful mechanical property that the generator will increase or decrease its speed slightly if the torque varies. This means that there will be less tear and wear on the gearbox due to lower peak torque. This is one of the most important reasons for using an asynchronous generator rather than a synchronous generator on a wind turbine which is directly connected to the electrical grid. The rotation of such direct grid connected asynchronous generators varies in the range of 100% of synchronous speed to 108% as shown in Fig. 5.10a.

In the section about the synchronous generator it has been discussed that it could run as a generator without connection to the public grid. An asynchronous generator is different, because it requires the stator to be magnetized from the grid before it works. An asynchronous generator can also run in a stand alone system, however, if it is provided with capacitors, which supply the necessary magnetization current. It also requires that there is some residual magnetism in the rotor iron, i.e. some leftover magnetism when the turbine is started. Otherwise a battery and power electronics or a small diesel generator is required to start the system. But it cannot run with constant frequency because the frequency is, as mentioned before, dependent on the rpm, which is dependent upon the generated electric power. In addition, an asynchronous generator can provide a limited amount of reactive power to the grid.

As discussed in the previous section, the motor (or generator) slip in an asynchronous (induction) machine is usually very small for reasons of efficiency, so the rotational speed will vary by up to 8% between idle and full load. The slip, however, is a function of the resistance in the rotor windings of the generator. So, one way of varying the slip is to vary the resistance in the rotor. In this way one may increase the generator slip to e.g. 10%. This can be done by having a wound rotor, i.e. a rotor with copper wire windings which are connected in a star, and connected with external variable resistors, plus an electronic control system to operate the resistors. The connection has usually been done with brushes and slip rings, which is a clear drawback over the elegantly simple technical design of a cage wound rotor machine. It also introduces parts which wear down in the generator. The rotation of such an asynchronous generator varies in the range of 100–110% of synchronous speed as shown in Fig. 5.10b.

Modern machines nowadays have double excited asynchronous generators as shown in Fig. 5.10c. In this case the generator is equipped with a winding in rotor and is excited with AC having variable frequency. This AC is produced by a DC link which is fed by the grid. This type of generators has an extended slip and is able to operate with rotational speed in the range of 100–120% of the synchronous speed. A further advantage of such generators is that they can supply more reactive power to the grid as compared to the other two types discussed above.

5.8 Issue of Noise and Its Control

The loudness or strength of a noise or sound signal is measured with the unit "decibel" abbreviated as "dB". One decibel is considered the smallest difference in sound level that the human ear can distinguish. Created in the early days of telephony as a way to measure cable and equipment performance and named after Alexander Graham Bell, decibels (dB) are a relative measurement derived from two signal levels: a reference input level and an observed output level. If a reference input level is used, the unit is often written as dB(A). A decibel is the logarithm of the ratio of the two levels. This means that if the dB level is doubling, the loudness level would be increasing by 10 times. A whisper is about 20 dB. A normal conversation is typically from 60 to 70 dB, and a noisy factory from 90 to 100 dB. Loud thunder is approximately 110 dB, and 120 dB borders on the threshold of pain.

The energy in sound waves (and thus the sound intensity) will drop with the square of the distance to the sound source, which is popularly known as the inverse square law. In other words, if one moves 200 m away from a wind turbine, the sound level will generally be one quarter of what it is 100 m away. To give an example: At one rotor diameter distance (43 m) from the base of a wind turbine emitting 100 dB(A) there will be generally a sound level of 55–60 dB(A) like a conversation. Four rotor diameters (170 m) away there will be 44 dB(A), corresponding to a quiet living room in a house. At 6 rotor diameters (260 m) it would be some 40 dB(A).

Sound emissions from wind turbines may have two different origins: Mechanical noise and aerodynamic noise. These two are explained below.

Mechanical noise, i.e. metal components moving or knocking against each other may originate in the gearbox, in the drive train (the shafts), and in the generator of a wind turbine.

When going by car, plane, or train, one may experience how resonance of different components, e.g. in the dashboard of a car or a window of a train may amplify noise. An important consideration, which enters into the turbine design process today, is the fact that the rotor blades may act as membranes that may retransmit noise vibrations from the nacelle and tower.

While designing, with the help of computer models it is ensured that the vibrations of different components do not interact to amplify noise. For example the chassis frame of the nacelle, on some of the wind turbines today, has some odd holes which are drilled into the chassis frame for no apparent reason. These holes were precisely made to ensure that the frame will not vibrate in step with the other components in the turbine.

When the wind hits different objects at a certain speed, it will generally start making a sound. If it hits the leaves of trees and bushes, or a water surface it will create a random mixture of high frequencies, often called white noise. The wind may also set surfaces in vibration, as sometimes happens with parts of a building, a car or even an engine-less glider plane. These surfaces in turn emit their own sound. If the wind hits a sharp edge, it may produce a pure tone.

In case of wind turbines, wind hits the rotor blades. Rotor blades make a slight swishing sound which may be heard close to a wind turbine at relatively low wind speeds. For their operation, rotor blades must split the wind on both sides of the blade so that energy can be transferred to the rotor. In the process they cause some emission of white noise. If the surfaces of the rotor blades are very smooth (which indeed they must be for aerodynamic reasons), the surfaces will emit a minor part of the noise. Most of the noise will originate from the trailing (back) edge of the blades. Careful design of trailing edges and very careful handling of rotor blades while they are mounted, have become routine practice in the industry for the purpose of controlling aerodynamic noise. Other things being equal, sound pressure will increase with the sixth power of the speed of the blade relative to the surrounding air. It can, therefore, be noticed that modern wind turbines with large rotor diameters have a very low rotational speed.

Since the tip of the blade moves substantially faster than the root of the blade, great care is taken about the design of the rotor tip. On close look at different rotor blades, it can be discovered that there has been substantial change in their geometry over time, as more and more research in the area has progressed. The research is also done for performance reasons, since most of the torque (rotational moment) of the rotor comes from the outer part of the blades. In addition, the airflows around the tip of rotor blades are extremely complex, compared to the airflow over the rest of the rotor blade.

One simple way of controlling noise is to reduce the tip-speed ratio by designing the rotor accordingly. An extension of this approach is sometimes used in Germany, by using different rotational speeds in day and night time due to different maximum permissible levels of noise. Many turbines can be seen with specially designed winglets at the tip of blades that help in reducing the noise, too, by controlling the turbulence at the tip.

Chapter 6
Operation and Control of Wind Energy Converters

After having discussed the appropriate wind energy system design, it is equally important to discuss its optimal operation and control of performance. The present chapter covers such aspects.

6.1 Power Curve and Capacity Factor

6.1.1 Power Curve

The power curve of a wind turbine is a graph that indicates which electrical power output will be available at different wind speeds. Figure 6.1 shows the shape of a theoretical power curve of a wind turbine.

Below the main features of any power curve are described:

(a) *The cut in wind speed* (v_c). Usually, wind turbines are designed to start running at wind speeds somewhere around 3–5 m/s. This is called the cut in wind speed. Below this speed of wind, the energy in wind is not sufficient to overcome the inertia of the rotor; hence, the machine does not produce any power below this speed of wind.

(b) *The cut out wind speed* (v_f). The wind turbine will be programmed to stop at high wind speeds above, say, 25 m/s, in order to avoid damaging the turbine or its surroundings. The stop wind speed is called the cut out wind speed.

Fig. 6.1 Theoretical power curve of a wind turbine

H.-J. Wagner, J. Mathur, *Introduction to Wind Energy Systems*, Green Energy and Technology, DOI 10.1007/978-3-642-02023-0_6, © Springer-Verlag Berlin Heidelberg 2009

(c) *Rated wind speed (v_r)*. The "rated wind speed" is the wind speed at which the "rated power" is achieved. This value for megawatt size turbines is about 12–15 m/s, and it corresponds to the point at which the conversion efficiency is near its maximum. The power output above the rated wind speed is mechanically or electrically maintained at a constant level, because the high output would destroy the equipment.

(d) *Survival speed*. Besides the above three important speeds of any power curve, there is one more speed which is specific together with the power curve: survival speed. It is that minimum speed at which the wind converter would not be able to sustain for its survival. The survival speed being more than cut-off speed, is not shown in the power curve. Its value is around 50–60 m/s. This value becomes one very important factor when selecting a wind turbine. It should be ensured that the maximum ever wind velocity in that location is lower than the survival speed of the machine.

It can be noted from the power curve that at lower wind speeds, the power output drops off sharply. This can be explained by the cubic power law, which states that the power available in the wind increases eight times for every doubling of wind speed and decreases eight times for every halving of the wind speed, as discussed in earlier chapters.

Using the power curve, it is possible to determine roughly how much power will be produced at the average or mean wind speed prevalent at a site. However, it is recommended to use the Weibull distribution for estimating the power output in connection with the power curve.

Power curves are calculated and found by field measurements, where an anemometer is placed on a mast reasonably close to the wind turbine (not on the turbine itself or too close to it, since the turbine rotor may create turbulence, and make wind speed measurement unreliable). If the wind speed is not fluctuating too rapidly, then one may use the wind speed measurements from the anemometer and read the electrical power output from the wind turbine and plot the two values together in a graph. In practice the wind speed always fluctuates, and one cannot measure exactly the column of wind that passes through the rotor of the turbine. It is not a workable solution just to place an anemometer in front of the turbine, since the turbine will also cast a "wind shadow" and brake the wind in front of itself.

In practice, therefore, one has to take an average of the different measurements for each wind speed, and plot the graph through these averages. Furthermore, it is difficult to make exact measurements of the wind speed itself. If one has a 3% error in wind speed measurement, then the energy in the wind may be 9% higher or lower (remember that the energy content varies with the third power of the wind speed). Consequently, there may be some errors in certified power curves. Nevertheless, a power curve is a very useful guide for estimating the output of a wind turbine.

6.1.2 Capacity Factor

Capacity factor is one element in measuring the productivity of a wind turbine or any other power production facility and to compare different locations with each other. It compares the plant's actual production over a given period of time, e.g. one year, with the amount of power the plant would have produced if it had run at full capacity for the same amount of time.

A conventional utility power plant, unless used as a peak load power plant, will normally run during most of the time unless it is idle only due to equipment problems or for maintenance or due to a reduced demand of energy. A capacity factor of 40% to 80% is typical for conventional plants. A wind plant is "fueled" by the wind, which does not blow steadily. Although modern utility-scale wind turbines typically operate 65% to 90% of the time, they often run at less than full capacity. Therefore, a capacity factor of 20% to 40% is common over one year, although they may achieve higher capacity factors during windy weeks or months and lower capacity factors in windless duration.

In Germany, another term, i.e. "load duration" is used to indicate the capacity factor. The term is a multiple of the total number of hours in one year (8,760 h) and the capacity factor. Since the capacity factor is much lower than one, the load duration also comes out to be much less than 8,760. For example, if the capacity factor is 20%, the load duration would be 8,760 × 0.2, which is 1,752 h.

The significance of load duration is that it expresses that number of hours for which the wind turbine can be considered to be virtually operating at its rated capacity in one year.

Here it is important to distinguish between plant availability factor and plant capacity factor. The first term refers to a fraction of one complete year for which the plant is available for use, irrespective of the availability of wind. For example, if a plant is under maintenance for 200 h in one year, out of a total of 8,760 h of one year, the availability factor would then be: (8760–200)/8760, i.e. 0.9771 or 97.71%.

To understand the capacity factor, let us consider a case of one 2 MW wind turbine:

In one year, a 2 MW wind turbine (8,760 h) can theoretically produce energy:
8,760h×2 MW = 17,520 MWh

If it actually produced 4,000 MWh only due to fluctuations in wind availability, the ratio of these two would be the capacity factor calculated below: Capacity factor = 4000/17520 = 0.2283 or 22.83%

It is important to note that while the capacity factor is almost entirely a matter of reliability for a fueled power plant, this is not the case with a wind plant. In case of a wind plant, it is a matter of economical turbine design. With a very large rotor and a very small generator, a wind turbine would run at full capacity whenever the wind blew and would have a 60–80% capacity factor, but it would produce very little electricity. The most electricity per unit of investment is gained by using a larger generator and accepting the fact that the capacity factor will be lower as a result. Wind turbines are fundamentally different from fueled power plants in this respect due to a limited and fluctuating availability of wind.

If a wind turbine's capacity factor is 33%, this does not mean it is only running one-third of the time. Rather, a wind turbine at a typical location would normally run for about 65–90% of the time. However, much of the time it will be generating at less than full capacity, making its capacity factor lower.

6.2 Power Control of Wind Turbines

Wind turbines are designed and operated to produce electrical energy as cheaply as possible. Wind turbines are, therefore, generally designed such that they yield a rated power output at wind speeds around 12–15 m/s. As discussed in previous sections, the exact value of such a wind velocity varies with manufacturer and size due to the difference in power curve. Its does not pay to design turbines that maximize their output at stronger winds, as such strong winds are rare. In case of stronger winds it is necessary to waste part of the excess energy of the wind in order to avoid damage to the wind turbine. Figure 6.2 shows the decline power coefficient for wind velocities exceeding the rated velocity.

Figure 6.2 shows the power curve of a 2.5 MW machine together with two more important curves: input power of wind together with the curve of multiplication of mechanical efficiency (η) and power coefficient (Cp). The product of these two represents the net efficiency of the wind turbine. The deviation between the input power and the power curve is due to net efficiency "$\eta \times c_p$". It can be very well observed that the curve for net efficiency reduces after reaching its maximum slightly before the rated wind speed. It is due to the fact that the pitch control starts wasting energy of wind after this point maintaining power under limits.

All wind turbines are designed with some sort of power control. There are two different ways of doing this safely on modern wind turbines.

6.2.1 Pitch Control

On a pitch controlled wind turbine the turbine's electronic controller checks the power output of the turbine several times per second. When the power output becomes too high, it sends an order to the blade pitch mechanism which immediately pitches (turns) the rotor blades slightly out of the wind. Conversely, the blades

Fig. 6.2 Capacity utilization and power curve of a wind turbine

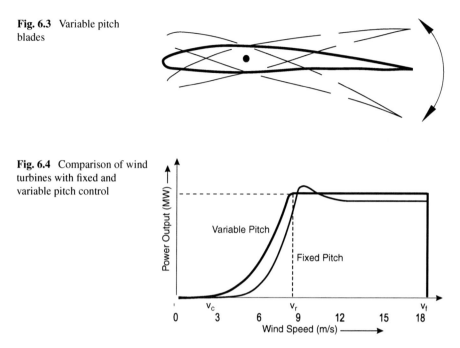

Fig. 6.3 Variable pitch blades

Fig. 6.4 Comparison of wind turbines with fixed and variable pitch control

are turned back into the wind whenever the wind drops again. The rotor blades thus have to be able to turn around their longitudinal axis (to pitch) as shown in Fig. 6.3.

During normal operation, the blades will pitch by fraction of a degree at a time - and the rotor will be turning at the same time. The pitch mechanism can be operated using hydraulic systems. But in most cases, individual electric drives are used to actuate control of blades, and the same mechanism is also used for applying brakes to the rotor by turning just one or two blades.

The importance of pitch control is clearly visible in Fig. 6.4. In the fixed pitch machine, power output dips very quickly at a wind speed higher than the rated wind speed and is oscillating around the rated power by higher wind speeds.. This can be understood in connection with Fig. 3.5 in Chap. 3, showing the relationship between power coefficient and tip speed ratio. As the wind speed increases, for a fixed rate of revolution of the rotor, the tip speed ratio decreases. As a result, the power coefficient decreases very sharply, which reduces the power output from the wind mill sharply as well. Whereas, in pitch controlled machines, due to a change of the blade angle, the curve itself changes which avoids a sudden dip in the power coefficient and output power.

6.2.1.1 Running a Pitch Controlled Turbine at Variable Speed

There are a number of advantages of being able to run a wind turbine at variable speed. One good reason for wanting to be able to run a turbine partially at variable speed is the fact that pitch-control (controlling the torque in order not to overload the gearbox and generator by pitching the wind turbine blades) is a mechanical process. This means that the reaction time for the pitch mechanism becomes an important factor in turbine design.

In a variable slip generator, however, one may start increasing its slip once it is close to the rated power of the turbine. The control strategy is to run the generator at half of its maximum slip when the turbine is operating near the rated power. When a wind gust occurs, the control mechanism signals to increase the generator slip to allow the rotor to run a bit faster, while the pitch mechanism begins to cope with the situation by pitching the blades more out of the wind. Once the pitch mechanism has done its work, the slip is decreased again. In case the wind suddenly drops, the process is applied in reverse.

6.2.2 Stall Control

An aircraft wing will stall, if the shape of the wing tapers off too quickly as the air moves along its general direction of motion. The turbulence is created on the back side of the wing in relation to the air current. Stall can be provoked if the surface of the aircraft wing – or the wind turbine rotor blade – is not completely even and smooth. A dent in the wing or rotor blade, or a piece like a self-adhesive tape can be enough to start the turbulence, even if the angle of the wing in relation to the general direction of airflow is small. Aircraft designers obviously try to avoid stall at all costs, since an aero plane without the lift from its wings will start falling down. Similarly, stall condition arriving in a wind energy converter would result in a no output condition.

In order to avoid stall condition, the rotor blades for large wind turbines are always twisted. Seen from the rotor blade, the wind will be coming from a much steeper angle (more from the general wind direction in the landscape), as one moves towards the root of the blade, and the centre of the rotor. As discussed above for stall, a rotor blade will stop giving lift, if the blade is hit at an angle of attack which is too steep. This is an additional reason for twisting blades to achieve an optimal angle of attack throughout the length of the blade. However, in the case of stall controlled wind turbines in particular, it is important that the blade is built such that it will stall gradually from the blade root and outwards at high wind speeds.

6.2.2.1 Passive Stall Control

Passive stall controlled wind turbines have the rotor blades bolted onto the hub at a fixed angle. The geometry of the rotor blade profile, however, has been aerodynamically designed to ensure that at the moment the wind speed becomes too high, it creates turbulence on the side of the rotor blade which is not facing the wind. This stall prevents the lifting force of the rotor blade from acting on the rotor.

The basic advantage of stall control over pitch control is that one avoids moving parts in the rotor itself, and a complex control system. On the other hand, stall control represents a very complex aerodynamic design problem, and related design challenges in the structural dynamics of the whole wind turbine, e.g. to avoid stall-induced vibrations. A normal passive stall controlled wind turbine will usually have a drop in the electrical power output for higher wind speeds, as the rotor blades go into deeper stall. Therefore, large wind machines are not using stall control for controlling output power.

6.2.2.2 Active Stall Control

An increasing number of larger wind turbines (1 MW and more) are being developed with an active stall power control mechanism. Technically the active stall machines resemble pitch controlled machines, since they have pitchable blades. In order to get a reasonably large torque (turning force) at low wind speeds, the machines will usually be programmed to pitch their blades much like a pitch controlled machine at low wind speeds. (Often they use only a few fixed steps depending upon the wind speed). When the machine reaches its rated power, however, an important difference from the pitch controlled machines can be noticed. The difference is that if the generator is about to be overloaded, the machine will pitch its blades in the opposite direction from what a pitch controlled machine does. In other words, it will increase the angle of attack of the rotor blades in order to make the blades go into a deeper stall, thus wasting the excess energy in the wind.

One of the advantages of active stall is that one can control the power output more accurately than with passive stall, so as to avoid overshooting the rated power of the machine at the beginning of a gust of wind. Another advantage is that the machine can be run almost exactly at rated power at all high wind speeds.

6.2.3 The Yaw Control

Some older wind turbines use flaps to control the power of the rotor, just like air-crafts use flaps to alter the geometry of the wings to provide extra lift at the time of takeoff. Another theoretical possibility is to yaw the rotor partly out of the wind to decrease power. This technique of yaw control is in practice used only for tiny wind turbines, as it subjects the rotor to cyclically varying stress which may ultimately damage the entire structure. The wind turbine yaw mechanism is used to turn the wind turbine rotor against the wind.

The wind turbine is said to have a yaw error, if the rotor is not perpendicular to the wind. A yaw error implies that a lower share of the energy in the wind will be running through the rotor area. The share will drop to the cosine of the yaw error, due to the change in the projected area of the rotor facing the wind.

If this were the only thing that happened, then yaw control would be an excellent way of controlling the power input to the wind turbine rotor. The main problem with yaw control is that the part of the rotor which is closest to the source direction of the wind, however, will be subject to a larger force (bending torque) than the rest of the rotor. On the one hand, this means that the rotor will have a tendency to yaw against the wind automatically, regardless of whether we are dealing with an upwind or a downwind turbine. On the other hand, it means that the blades will be bending back and forth in a flapwise direction for each turn of the rotor. Wind turbines which are running with a yaw error are therefore subject to larger fatigue loads than wind turbines which are yawed in a perpendicular direction against the wind.

6.3 Connection to the Grid

The connection to grid depends upon the purpose for which a wind energy system is used. Although most large wind energy systems are installed to feed electricity to the grid, however, other configurations are also found, particularly in small and medium size turbines. Their requirements for grid connection are described in this section.

6.3.1 Applications of Wind Energy Converters

From the point of view of utilization of generated electricity, wind energy systems can be used in three ways as explained below:

(a) *Grid connected system.* The grid connected system has a connection with an electricity transmission and distribution system called grid-connected systems. A grid-connected wind turbine can be used for reducing the consumption of fossil energy carriers. If the turbine cannot deliver the amount of energy needed, the public utility makes up the difference.

(b) *Stand alone system.* As their name suggests, stand-alone systems are not connected to the utility grid. Stand-alone wind energy systems can be appropriate for homes, farms, or even entire communities (a co-housing project, for example) that are far from the nearest utility lines. However, if grid is available in a nearby area, it is always advantageous to be connected to the grid. Therefore, stand alone systems are used only in areas where grid is not available at all.

(c) *Hybrid systems.* A hybrid system using wind energy is one in which operation of a wind turbine is combined with any other source of power. Such a combination can be considered as another type of stand alone system. This source may be e.g. a photovoltaic system, or a diesel generator set. To specify the nature of these systems, sometimes the term "stand-alone hybrid system" is also used to differentiate between the hybrid systems in which wind turbine is operating in hybrid mode with the grid supply.

At many sites, wind speeds are low in the summer when the sun shines brightest and longest. The wind is strong in the winter when there is less sunlight available. Since the peak operating times for wind and photovoltaic systems occur at different times of the day and year, hybrid systems are more likely to produce power when required.

For the times when neither the wind generator nor the photovoltaic modules are producing electricity (for example, at night when the wind is not blowing), most stand-alone systems provide power through batteries and/or an engine-generator powered by fossil fuels like diesel.

If the batteries run low, the engine-generator can be run at full power until the batteries are charged. Adding a fossil-fuel-powered generator makes the system more complex, but modern electronic controllers can operate these complex systems automatically. Adding an engine-generator can also reduce the number of photovoltaic modules and batteries in the system. It should be ensured that the storage capability is large enough to supply electrical needs during non-charging periods. Battery banks are typically sized for one to three days of windless operation.

6.3.2 Voltage Requirement

A single small converter can be directly connected into the grid network at e.g. 0.4 kV level. Once the wind energy converter is integrated into the grid network, there must be very limited voltage change, voltage oscillation or flicker experienced in the homes on that network branch. The loss of voltage due to resistance in the cabling can be avoided by increasing the diameter of the cables. It is often required that a new network branch is constructed and linked to the transformer in order to reduce the voltage disturbances. This increases the installation costs of the converter.

A single Mega-Watt size converter cannot be connected to the grid at the 0.4 kV stage, but has to be connected at 10–30 kV, which is the usual level of the city electricity share distribution. In remote areas, where a 30 kV connection is not established, the connection must be created and financed. Wind parks with a lot of Mega-Watt converters must be connected into the electrical grid at a level of about 100 kV and higher.

As mentioned earlier, the maximum power output is obtained only in a few hours during the year. Figure 6.5 shows a typical load distribution, measured within the German 250 MW program. With larger wind energy installations, this uneven distribution leads to the need of higher regulation capacities by conventional power systems in the future.

It would be worth mentioning here that due to fluctuations in availability of wind, these systems alone cannot fully meet the energy demand. On a large scale, additional conventional power plants that are flexible in their level of operation are needed. Whenever wind is available, the level of operation of the conventional power plant is reduced, and when wind is not available, it is increased to meet the demand.

6.3.3 Special Aspects of the Connection of Offshore Wind Parks

Grid connection of offshore wind farms is a technical and economical challenge to both wind turbine and grid operators. In the initial phase, the still quite limited capacity of early pilot farms allows the use of a conventional three-phase AC

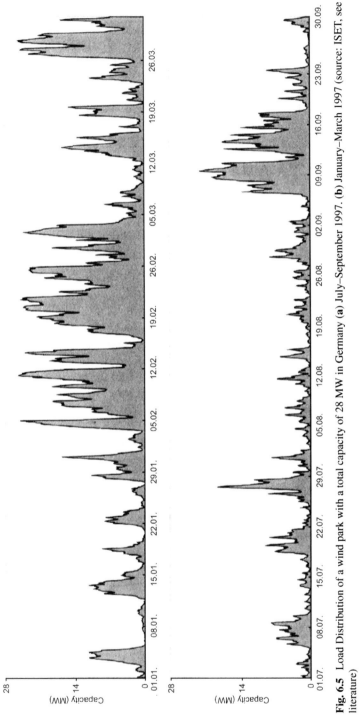

Fig. 6.5 Load Distribution of a wind park with a total capacity of 28 MW in Germany (**a**) July–September 1997. (**b**) January–March 1997 (source: ISET, see literature)

connection to the onshore grid system which is a well known and inexpensive technology.

An internal grid is necessary to connect the offshore wind farm to the onshore grid. The produced power has to be fed to an offshore transformer substation, to which wind turbines are connected via undersea cables by a voltage of e.g. 30 kV. After stepping-up to the transmission line voltage, the power is conveyed to the shore.

Better suitability of submarine High Voltage DC transmission lines for offshore wind turbines is still a debatable issue. High Voltage DC lines offer the advantage of reduced transmission losses but they have the problem of higher cost. The issue is much more important as compared to on-shore turbines due to greater lengths of transmission, which are of the order of 5–10 km.

Chapter 7
Economics and Policy Issues

This chapter presents various economic and financial considerations that are important for selecting wind turbines, and before taking a decision in favor of installing a wind turbine at all.

7.1 Cost of Wind Turbines

7.1.1 Initial Cost of Wind Turbine

The average price of large, modern onshore wind farms in Germany is around 1,250 € per kilowatt electrical power installed, in India it is slightly less than 1,000 € (65,000 Indian Rupees) per kilowatt, mainly due to cheaper labor costs. These costs are just indicative figures and would differ with capacities and make. For single turbines or small clusters of turbines the costs will usually be somewhat higher as compared to large wind farms, due to bulk purchase price breaks. These costs also increase with an increase in tower height.

Installation costs include foundations, road construction (necessary to move the turbine and the sections of the tower to the building site), a transformer (necessary to convert the low voltage e.g. 690 V) current from the turbine to 10–30 kV current for the local electrical grid, telephone connection for remote control and surveillance of the turbine, and cabling costs, i.e. the cable from the turbine to the local 10–30 kV power line.

Obviously, the costs of roads and foundations depend on soil conditions, i.e. how cheap and easy it is to build a road capable of carrying heavy trucks. Another variable factor is the distance to the nearest ordinary road, the cost of getting a mobile crane to the site, and the distance to a power line capable of handling the maximum energy output from the turbine. Transportation costs for the turbine may enter the calculation, if the site is very remote. Of course this figure also depends upon the weight of the turbine, distance to be covered and, last but not least, current fuel prices.

It is obviously cheaper to connect many turbines at the same location, than just one. On the other hand, there are limits to the amount of electrical energy the local

H.-J. Wagner, J. Mathur, *Introduction to Wind Energy Systems*, Green Energy and Technology, DOI 10.1007/978-3-642-02023-0_7, © Springer-Verlag Berlin Heidelberg 2009

electrical grid can handle. Especially in developing countries like India, if the local grid is too weak to handle the output from the turbine, a need for grid reinforcement may arise, i.e. by extending the high voltage electrical grid. Agency that bears the cost of grid reinforcement: the power company or the turbine owner; varies from country to country.

7.1.2 Operation and Maintenance Costs for Wind Turbines

Modern wind turbines are designed to work for some 120,000 h of operation throughout their design lifetime of 20 years.

7.1.2.1 Operation Maintenance Costs

Experience shows that maintenance costs are generally very low when the turbines are brand new, but they increase somewhat as the turbines age. The newer generations of turbines have relatively lower repair and maintenance costs than the older generations. Older wind turbines (25–150 kW) require annual maintenance costs at an average of around 3% of the original turbine investment. Newer turbines are on average substantially larger, which would tend to lower maintenance costs per kW installed power, since it is not needed to service a large, modern machine more often than a small one. For newer machines the estimates range around 1.5–2% per year of the original turbine investment. In some cases, however, e.g. if the product is not standardized and/or the site is not investigated thoroughly, even these costs may be relatively high as compared to industry average.

Most of the maintenances cost are expressed as a fixed amount per year for the regular service of the turbines, but some people prefer to use a fixed amount per kWh of output in their calculations, for example around 0.01 €/kWh. The reasoning behind this method is that tear and wear on the turbine generally increases with increasing production.

In addition to maintenance, another important cost element is insurance costs. Insurance of wind turbines is required to secure the loss of their heavy investment (fully or partially as damage of parts) due to any unforeseen damage, such as lightning or hurricanes, or any other factor.

Other than the economies of scale which vary with the size of the turbine, as mentioned above, there may be economies of scale in the operation of wind parks rather than individual turbines. These economies are related to the semi-annual maintenance visits, surveillance and administration, etc.

7.1.2.2 Turbine Reinvestment (Refurbishment, Major Overhauls)

Some wind turbine components are more subject to tear and wear than others. This is particularly true for rotor blades and gearboxes. Wind turbine owners who see that their turbine is close the end of their technical design lifetime may find it advantageous to increase the lifetime of the turbine by doing a major overhaul of the turbine, e.g. by replacing the rotor blades after 7–10 years. The price of a new set of rotor blades, a gearbox, or a generator is usually in the order of magnitude of 15–20% of the price of the turbine.

7.1.2.3 The Availability Factor

The figures for annual energy output assume that wind turbines are operational and ready to run all the time. In practice, however, wind turbines need servicing and regular inspection (e.g. once every six months) to ensure that they remain safe. In addition, component failures and accidents (such as lightning strikes) may disable wind turbines. The best turbine manufacturers consistently achieve availability factors up to 98%, i.e. the machines offer technical availability of up to 98% of the time. Total energy output is generally affected by less than 2%, since wind turbines are never serviced during high winds. The availability factor is, therefore, usually ignored when doing economic calculations, since other uncertainties (e.g. wind variability) are far larger.

It is always a good idea to check the manufacturers' track record and servicing ability before buying a new wind turbine. Manufacturers in most cases offer a guarantee of availability of the machine for the initial 2–3 years. Afterwards, they prefer to enter into a sort of annual maintenance contract with the machine owner, in which they charge a fixed amount to guarantee a certain minimum availability of the system for the generation of power.

7.2 Electrical Tariffs

Five different types of tariff systems are found in connection with wind energy systems due to a difference in policies which change in country/location as described below:

(a) *Feed-in tariff*. The feed-in tariff scheme, as its name suggests, is based upon the principle of paying an amount to the wind energy producer that depends on the amount of electricity fed into the grid. This is done at a pre-declared rate per unit of electricity. This rate is higher than the rate of production of electricity from a conventional (using fossil or nuclear fuels) power plant. The most important aspect in a feed-in tariff system is that the grid cannot deny accepting the power generated by the wind energy system, even if it is surplus. The feed-in tariff system exists e.g. in Germany as well as in many other countries. The rates of feed-in tariff change with respect to the location within the country. To promote the use of wind energy systems by improving their economics, currently, in Germany, the feed-in-tariff rates are higher in inland locations as compared to coastal locations which are attractive for wind energy anyway.

If an energy producer owning wind turbines produces a certain amount of electricity and consumes the same amount of electricity, separate accounting is done for the electricity produced by the windmill and the electricity consumed. The feed-in tariff rate is not uniform for all renewable energy systems, e.g. in the case of photovoltaic systems, the feed-in tariff rates are even higher than those of wind energy systems.

Feed-in tariff rates also exist in India, but a difference in feed-in tariff rates between coastal and non-coastal areas does not exist. Currently, the rates are the same over one particular state; they are the highest in the state of Maharashtra and the lowest in the state of Tamilnadu.

(b) *Net metering system.* Net metering or net billing is a term applied to laws and programs under which a utility allows the meter of a customer with a wind turbine, to turn backward, thereby, in effect, allowing the customer to deliver any excess electricity he produces to the utility and be credited on a one-for-one basis against any electricity the utility supplies to him. For example: During a one-month period, one wind turbine in a farm house generates 300 kilowatt-hours (kWh) of electricity. Most of the electricity is generated at a time when the equipment in the farm house (refrigerator, lights, etc.) draws electricity, and the power generated is used on site. However, some is generated at night when most equipment is turned off. At the end of the month, the turbine has generated 100 kWh in excess of local needs, and that electricity has been transmitted to the utility system. Over the same one-month period, the utility system supplied to the farm house electricity of 500 kWh for use at times when the wind turbine was not or insufficiently generating. Since the meter ran backward while the 100 kWh was being transmitted to the utility system, the farm house will be billed for 400 kWh rather than 500 kWh.

The net-metering scheme can improve the economics of captive wind turbine by allowing the turbine's owner to use the excess electricity to offset utility-supplied power at the full retail rate, rather than having to sell the power to the utility at the price the utility pays for the wholesale electricity it buys or generates itself. Many utilities argue against net metering laws, saying that they are being required, in effect, to buy power from wind turbine owners at full retail rates, and are, therefore, being deprived of a profit from part of their electricity sales. However, wind energy advocates have successfully argued that what is going on is a power swap, and that it is standard practice in the utility industry for utilities to trade power among themselves without accounting for differences in the costs of generating the various kilowatt-hours involved.

Although in Europe the feed-in tariff scheme exists, the popularity of net metering is growing, worldwide. Some countries like India even offer limited flexibility in the installation of wind turbines and consumption centers. This means that they may not be exactly at the same place. In such cases, of course, the utility charges a nominal amount termed as "wheeling charges" against use of their infrastructure/power lines for transferring power from the point of generation to the point of consumption. In such arrangements, there are, of course, some limitations such as geographical and commercial boundaries which differ from state to state.

(c) *Time dependent rates.* Ideally, electricity companies are more interested in buying electricity during the periods of peak load (maximum consumption) on the electrical grid. Therefore, in some areas, power companies apply variable electricity tariffs depending on the time of day, when they buy electrical energy from private wind turbine owners. Normally, wind turbine owners receive less than the normal consumer price of electricity, since that price usually includes payment for the power company's operation and maintenance of the electrical grid, plus its profits. In locations where a feed-in tariff system exists, as in Germany, the time when power is generated becomes non-significant.

(d) *Quota system.* An additional system is that of the allocation of "quota" of renewable energy. In this system, every producer of electricity for grid is given a "quota", e.g. 20%. In the total electricity produced by every company, there has to be 20% share of energy coming from renewable sources such as wind and solar. In case this quota is not met, provisions of penalties are made. The European Union is discussing to adopt this system in Europe. Provisions are being developed that if a company produces more energy from renewables than its quota, e.g. 25%, it would be granted a certificate for this excess renewable energy. This certificate can be purchased by other companies that are having a lower share of renewables from the quota, e.g. 15%. By purchasing these certificates, the second company will also be considered to comply with its quota. This system is quite similar to the concept of "emission trading" for greenhouse gases. Up to the middle of 2008, there was no consensus on bringing such a quota system to Europe. The argument was debated that with such a quota system, relatively costlier renewable energy technologies, such as solar photovoltaic systems, would not be installed at all. Companies might prefer to purchase certificates in place of installing PV systems. The concept of the quota system is still being discussed at EU level.

(e) *Production tax incentives/Investment incentives.* The Production tax incentive is a generation-based mechanism, which supports renewable energies through payment exemptions from electricity taxes, e.g. the energy tax for renewable energies, applied to all producers. Hence it is a system that affords an avoided cost on the producer side. Also the Investment incentive is a mechanism, like the name leads one to divine, to lower the costs for the investment in renewable energies so that it gets more attractive funding.

7.3 Mechanisms to Support Funding

7.3.1 Capacity Credit

To understand the concept of the capacity credit, we look at its opposite, the power tariffs: Large electricity customers are usually charged both for the amount of energy (kWh) they use, and for the maximum amount of power (kW) they draw from the grid, i.e. customers having more connected load drawing a lot of energy very quickly have to pay more. The reason they have to pay more is that this fact obliges the power company to have a higher total generating capacity (more power plant) available. Power companies have to consider adding generating capacity whenever they give new consumers access to the grid. But with a modest number of wind turbines in the grid, wind turbines are almost like "negative consumers", since they postpone the need to install other new generating capacity. Many power companies in industrialized countries, therefore, pay a certain amount per year to the wind turbine owner as a capacity credit. The exact level of the capacity credit varies with location benefit to utility in terms of avoided installation of power plants. In some countries it is paid on the basis of a number of measurements of power output during the year. In other areas, some other formula is used. In a number of countries like Germany, no capacity credit is given, as it is assumed to be part of the energy tariff due to the fact that due to fluctuating output from wind energy system, the utility anyways has to install power plants as back up for a no wind situation.

7.3.2 Environmental Credit and Clear Development Mechanism

Many governments and power companies around the world wish to promote the use of renewable energy sources. Therefore, in developed or industrial countries, they offer a certain environmental premium to wind energy, e.g. in the form of a refund of electricity taxes etc. on top of normal rates paid for electricity delivered to the grid. In developing countries like India, due to the fact that every unit of electricity generated avoids generation of the same amount of electricity from fossil fuel based power plants, the avoided environmental emissions offer the opportunity for additional earnings through the Clean Development Mechanism (CDM) under the Kyoto Protocol. Every ton of carbondioxide gas saved or avoided, termed as one carbon credit, is sold to industrialized countries at rates of around 15 €. These additional earnings help to improve the economics of power generation, which has been one strong reason behind the rapid increase in the wind energy based installed capacity for power generation in India.

7.3.3 Tax Benefits

In some countries like India, several tax benefits are offered to promote wind energy systems. Major incentives or benefits are mentioned here:

(a) If the wind energy converter or its parts or components are imported from other countries, no excise duty (import tax) is applicable.
(b) Exemption/reduction in central government sales tax and general sales tax are available on the sale of renewable energy equipment in various states of India.
(c) 80% accelerated depreciation on specified renewable energy devices/systems (including wind power equipment) in the first few years of installation of the projects.
(d) There is an income tax holiday on the income generated from renewable energy systems including wind energy through power generation.

As a result of these measures, several investors have installed wind energy systems in different parts of India.

7.4 Wind Energy Economics

7.4.1 Financial Analysis – A Case Study for India

The annual electricity production will vary enormously depending on the amount of wind on the turbine site. Therefore, there is not a single price for wind energy. It differs from year to year and location to location, that means wind speed and wind profiles.

Following a case study from India should give an impression about the factors which has to be considered to calculate the electricity costs from wind turbines.

It is proposed to install a 1,250 kW wind power plant at a certain location where the capacity utilization for power generation over one year is expected to be 35%. Below are the major cost elements in Indian rupees (Rs.):

Currency: Rs. 100 = 1,54 €; 1 € = Rs. 65
Cost of land: Rs. 1 million
Cost of wind energy converter: Rs. 60 million
Erection and commissioning charges: Rs. 6 million
Sub station charges: Rs. 4.0 million

Operation and maintenance charges: Free during first year, second year Rs. 1.1 million, for subsequent years with an escalation of 5% per year.

Other economic parameters include a total generation as 3.85 million kWh electricity (at 35% capacity utilization), and losses as 7%.

Tariff for selling the electricity is Rs. 5.5/kWh in the first year, with a price escalation of 2% per year.

Wheeling charges: 5% of tariff

Life cycle term of the project is 20 years.

The investment requirement, power generation and sale of power can be summarized as in the Table below (Table 7.1):

Table 7.2 shows a detailed cash flow analysis of the project under various heads:

It can be seen from the above table that the cumulative cash accrual becomes equal to the initial investment of Rs. 71 million at the end of the fourth year. This period is known as the simple payback period of investment. Beyond this period and up to the end of project life (20 years in this case), every cash accrual is net profit from the project. If the financial discounting rate is significant and the payback period is longer, it is recommended to include the time based value of money for estimating the payback period.

7.4.2 Financing of a Wind Park – A Case in Germany

Due to significant differences in the amounts invested for an onshore and off-shore wind park, there are two different cases to describe. The amount invested

Table 7.1 Assumptions for the case study of financial analysis

Investment requirement	
Cost of land (million Rs.)	1
Cost of wind machine (million Rs.)	60
Erection and commissioning charges (million Rs.)	6
Sub-station charges (million Rs.)	4
Total cost (million Rs.)	71
Running cost requirement	
O & M (million Rs.) per year	1.1
Escalation	5%
Free O & M (Years)	1
Depreciation rate	80%
Insurance (million Rs./year)	0.15
Tariff (Rs./kWh)	5.5
Escalation in tariff per year	2%
Wheeling charges (% of tariff)	5%
Generation (million kWh)	3.85
Losses	7%
Sale of power	
Tariff (Rs./kWh)	5.5
Wheeling charges (5% of tariff)	0.28
Actual tariff (tariff – wheeling charges) (Rs./kWh)	5.23
Generation (million kWh/year)	3.85
Losses(7% of generation)	0.27
Actual generation (generation – losses) (million kWh/year)	3.58

for an e.g. 15 MW onshore wind park is about 19 million €, which is split in an equity capital of at least 4 million € and a dept capital of 15 million €. The wind power plant, the installation, the commissioning and the infrastructure constitutes

Table 7.2 Cash flow analysis of a wind energy project

Year	1	2	3	4	5	6–20
Receipt (million Rs.)						
Power rate (Rs./kWh)	5.23	5.28	5.33	5.38	5.44	Increasing as per rate of tariff
Receipt (sale of power)	18.71	18.90	19.08	19.27	19.47	Increasing from 6th year to 20th year
Total receipts (A)	18.71	18.90	19.08	19.27	19.47	Increasing from 6th year to 20th year
Expenses (million Rs.)						
O & M (free for 1st year)	0	1.1	1.155	1.21	1.27	Increasing from 6th year to 20th year
Insurance	0.15	0.15	0.15	0.15	0.15	0.15
Total expenses (B)	0.15	1.25	1.31	1.36	1.42	Increasing from 6th year to 20th year
Cumulative cash accrual (million Rs.) (A–B)	18.56	36.20	53.98	71.89	89.94	Increasing up to 20th year (project life)

an 83% proportion of the investment volume. With 11%, technical planning, surface investigation, a survey report as well as the grid connection costs also capture
a capacious amount of the investment volume. Compensatory measures, compen-
sations for use during the building-up phase, cost for the partnership during the
investment phase, the consulting and the financing during the construction period
adopt rather inferior parts.

   To raise the equity capital a limited partnership (L.P.) is founded. This L.P. incurs
the sale of the interests and covers the banker's guarantee. It is also responsible for
the construction, the instigation and the disposal of the produced electric power.
The limited partner's individual deposit constitutes the equity capital and starts from
several thousand Euros per partner. The bank allocates the dept capital in form of
a loan. This loan could be for example a loan of 11 million €, with a duration of
15 years and an interest loan of about 6%. As a chattel mortgage the wind park, the
feed-in tariffs and the liquidity reserves are transferred to the bank until the loan is
paid off. The money for the bank has priority over the payout of dividends for the
limited partners.

   The funding of an offshore wind park requires an investment volume of at least
200 million € and more. Therefore, a limited partnership cannot be established and
grand transmission providers have to take over this part. Due to the lack of experience of the instigation of offshore parks up to now it is difficult to find cooperative
banks to finance these parks, which retard the process of the installation of win
parks.

   In such a case it could be helpful if the government stand security for the
financing of the first windparks.

7.5 Wind Turbines After Operational Life

The amount of material used in wind turbines increase with the size and the number
of wind turbines. After the operational life of the equipment it must be demolated
and as far as possible recycled. Recycling of steel and other metals do not create
problems, whereas the rotor blades, made of plastic, glass fibers or carbon fibers are
at least difficult to recycle. Therefore industry should develop solutions to get rid of
the waste in ecological acceptable waste treatment produces.

Chapter 8
Outlook

More than 50 countries around the world have an organized set up in the field of wind power. Large progress in wind power has been witnessed in the countries of the European Union, but the situation is changing quite rapidly. The United States and Canada are both experiencing a surge of activity, while markets like Asia and South America are also catching up quite rapidly. In the past few years, a new frontier for wind power development has also been established in the sea. With offshore wind parks beginning to make a contribution, the possibilities of using wind energy systems worldwide are likely to increase many times in the future. Establishing wind energy projects in the sea has opened up new demands, including the need for stronger foundations, long underwater cables and larger individual turbines, but offshore wind parks are expected to contribute an increasing proportion of global capacity.

With its increase in the global market, wind power has seen a fall in cost. A modern wind turbine produces more electricity at less cost per unit (kWh) than its equivalent 20 years ago. At good locations wind may compete with, and even beat, the cost of both coal and gas-fired power, if the future carbon dioxide certificate costs are taken into account. If the "external costs" associated with the pollution and health effects resulting from fossil fuel and nuclear generation are fully taken into account, wind power gains an additional advantage.

Studies of different institutions around the world have confirmed that a lack of wind is not likely to be a limiting factor on global wind power development. With the expansion of the wind industry, large quantities of wind powered electricity will need to be integrated into the grid network. Because of the variability of the wind, control methods have to be established for dealing with variations in demand and supply. Their installation could help to handle this issue.

It has been estimated by the Global Wind Energy Council that wind power could supply one third of the world's electricity by 2050. This is a big challenge not only for the wind energy sector, but for the entire electricity supply system. Let's try to catch it.

H.-J. Wagner, J. Mathur, *Introduction to Wind Energy Systems*, Green Energy and Technology, DOI 10.1007/978-3-642-02023-0_8, © Springer-Verlag Berlin Heidelberg 2009

Glossary

Alternating Current (AC): An electric current that reverses its direction at regularly recurring intervals, usually 50 or 60 times per second. Today's grids are operated by Alternating Current.

Direct Current (DC): An electric current with constant direction. It can be used on high voltage level to transport bulk electricity with less losses.

Gigawatt (GW): A unit of power equal to 1 billion watts, 1 million kilowatts, or 1,000 megawatts.

Joule (J): A standard international unit of energy; 1,055 joules are equal to 1 BTU.

Kilowatt (kW): A standard unit of electrical power equal to 1,000 watts, or to the energy consumption at a rate of 1,000 joules per second.

Kilowatt-Hour (kWh): 1,000 watts acting over a period of 1 hour. The kWh is a unit of energy. 1 kWh = 3,600 kJ.

Megawatt (MW): 1,000 kilowatts, or 1 million watts, standard measure of electric power plant generating capacity.

Megawatt-Hour (MWh): 1,000 kilowatts-hours or 1 million watt-hours.

Watt (W): The rate of energy transfer equivalent to one ampere under an electrical pressure of one volt. One watt equals 1/746 horsepower, or one joule per second. It is the product of voltage and current (Amperage).

Watt-hour (Wh): The electrical energy unit of measure equal to 1 watt of power supplied to, or taken from, an electric circuit steadily for 1 hour.

H.-J. Wagner, J. Mathur, *Introduction to Wind Energy Systems*, Green Energy and Technology, DOI 10.1007/978-3-642-02023-0_BM2, © Springer-Verlag Berlin Heidelberg 2009

Literature

Annual report 2007–2008. Centre of Wind Energy Technology, Government of India, http://cwet.tn.nic.in/

Annual report 2007–2008. Indian Renewable Energy Development Corporation, http://www.ireda.in/incentives.asp

Annual report 2007–2008. Ministry of New and Renewable Energy sources, India, http://www.mnes.ni.in/

Bansal, N K, Kleemann, M, Meliss, M (1990) Renewable Energy Sources and Conversion Technology. Tata McGraw-Hill Publishing Company Ltd., New Delhi, ISBN 0-07-4600023-0

Global Wind 2008 Report. Global Wind Energy Council http://gwec.net

Heier, S (2006) Grid integration of wind energy conversion systems. John Wiley & Sons Ltd., ISBN 978-0470868997

Landolt-Börnstein (2006) Energy Technologies, Subvolume C Renewable Energy. Springer Verlag S. 233–241, ISBN 3-540-42962-X

Pantel Mukund, R (2006) Wind and Solar Power Systems – Design, Analysis and Operation. CRC Press, ISBN 0849315700

Scientific Measure and Evaluation Program for the 250 MW-Wind Test Program, Results of 1997 (1998). Wissenschaftliches Mess- und Evaluierungsprogramm (WMEP) zum Breitentest 250 MW Wind, Jahresauswertung im Auftrag des Bundesministeriums für Bildung, Wissenschaft, Forschung und Technologie, Kassel: Institut für Solare Energieversorgungstechnike n (ISET) (in German language)

Index

Breinigsville, PA USA
07 September 2010
244753BV00005B/1/P